RESEARCH METHODS

A Practical Guide for
Students and Researchers

RESEARCH METHODS

A Practical Guide for
Students and Researchers

Willie Tan

NUS, Singapore

NEW JERSEY · LONDON · SINGAPORE · BEIJING · SHANGHAI · HONG KONG · TAIPEI · CHENNAI · TOKYO

Published by

World Scientific Publishing Co. Pte. Ltd.
5 Toh Tuck Link, Singapore 596224
USA office: 27 Warren Street, Suite 401-402, Hackensack, NJ 07601
UK office: 57 Shelton Street, Covent Garden, London WC2H 9HE

National Library Board, Singapore Cataloguing in Publication Data
Names: Tan, Willie.
Title: Research methods : a practical guide for students and researchers /
    Willie Tan, NUS, Singapore.
Description: Singapore : World Scientific, [2017] | Includes bibliographic references and index.
Identifiers: OCN 992937200 | 978-981-32-2961-7 (paperback) | 978-981-32-2958-7 (hardcover)
Subjects: LCSH: Research--Methodology.
Classification: DDC 001.42--dc23

British Library Cataloguing-in-Publication Data
A catalogue record for this book is available from the British Library.

Desk Editor: Amanda Yun

Printed in Singapore

Preface

This book is based on a series of lectures given over many years to undergraduate and graduate students in the Department of Building at the School of Design and Environment, National University of Singapore. The students are largely from architecture, construction, engineering (all branches), information technology, and the service sector.

The target audience of this book includes senior undergraduates, graduate coursework students, researchers, and professionals. My aim in writing this book is to provide a *concise*, *practical*, and reasonably *comprehensive* guide on research methods. I hope it will help readers who are just starting out on their research journey. The book is also *integrative*, in the sense that it connects the different steps of the research process.

In terms of *conciseness*, most of the chapters are relatively short so that the reader can focus on the key issues in each step of the research process. I first explain the main principles before discussing their applications using relevant examples. Many of these examples are actual research conducted by experts in various disciplines. Hence, the book is suitable for readers from different fields.

The book has a *practical* slant. Many books on research methods are too simple and impractical as a guide to actual practice. These authors simplify the exposition to avoid the harder statistics. However, it also means that readers are missing a vital tool of research. I have tried to explain the statistics more intuitively so that readers can see the forest before they examine the trees in detail. I hope that this will make the learning journey more enjoyable and less of an obstacle course.

In terms of *comprehensiveness*, the book covers both qualitative and quantitative research designs; that is, it discusses the case study, comparative study, experiment, survey, and regression. The case study and comparative study are small-N research designs, where N stands for sample size. The experiment, survey, and regression studies are large-N research designs. I organized the research designs into two main categories, namely, causal and interpretive research.

Importantly, this book *integrates* the entire research process of problem formulation, hypothesis, research methodology, data analysis, and conclusion. For example, it does not provide a smorgasbord of techniques for analyzing data. Instead, I arranged the chapters on data analysis by research methodologies. In this way, we avoid the common problem of analyzing data using inappropriate tools.

I have also paid considerable attention to the difficulties students face in formulating the research problem, reviewing the literature, and developing the research framework or hypothesis. To motivate and guide students, I provided detailed examples from different disciplines on how to execute these steps effectively.

I would like to thank my research assistant, Zhang Yajian, for reading the entire draft and providing useful suggestions. Jonathan Lian, Eric Tan, Yeo Teng Kwong, Daniel Wong, Lim Pin, Gabriel Kor, Leonard Teo, Dilini Thoradeniya, and Calvin Yeung read draft chapters.

Lastly, I thank Ms. Amanda Yun, Senior Editor, World Scientific Publishing, for her kind assistance and patience while I seem to work endlessly on the draft because of my busy schedule.

Willie Tan

Contents

Appendix

CHAPTER 1
Introduction to Research

What is Science?

Research is the bedrock of science. Before discussing what constitutes scientific research, we first need to understand what constitutes science.

Most people learn science without knowing what science really entails. As early as primary school, the child learns chemistry, physics, and biology as subject matters in a "general science" course, topic by topic. The teacher emphasizes the content of science rather than the methodology. This way of approaching science continues in secondary and high schools as well as junior colleges at higher levels of difficulty. Gradually, "general science" splits into the various sciences such as biology, physics, and chemistry at higher levels.

At university, the Faculty of Science normally offers "pure" science subjects such as the mathematical sciences (for example, mathematics and statistics) and natural sciences (for example, physics and biology). At the Medical School, there are medical sciences (for example, physiology and pharmacology). There is also cognitive science, that deals with human perception, memory, learning, and reasoning. The applied sciences may be found at the Faculty of Engineering, and there is also computer science at the School of Computing. Finally, the Faculty of Social Science offers courses in economics, sociology, psychology, and political science.

The student discovers that, despite the varying content, all these sciences have something in common. They have similar aims, such as to explore, describe, interpret, explain, predict, control or evaluate phenomena or behaviors.

However, science is not, and cannot be, defined by its content or aims. If it is defined by content, there will be considerable disagreements as to what constitutes science. For example, what aspects of human behavior

1

should be considered science? Should it be restricted to physical behavior, or should it also include emotional behavior that are shaped by culture, authority, attitudes, values, beliefs, morals, and so on? Similarly, if science is defined by aims, then other non-scientific studies have similar aims. For example, one of the aims of astrology is to predict the future based on the belief that there is a close connection between astronomical phenomena and human events. By providing guidance on how we should think and act, astrologists hope that we can lead better lives.

If what distinguishes science from non-science cannot be the content or aims, then what is it? The answer, which is common to all sciences, is its *methodologies* or, more precisely, scientific methodologies. In other words, scientific research is a systematic and careful process of inquiry or investigation based on theory and evidence to seek new understandings and contribute to knowledge.

There are two main scientific methodologies, namely, causal and interpretative. Causal science seeks to identify causes and effects, such as the effect of changing money supply on general prices. Interpretive science seeks to uncover different human understandings, perceptions, or perspectives of the same event. For example, two parties in a project dispute may have different understandings of the issues.

Science as Voyage of Discovery

Science is sometimes described as a voyage of discovery into the unknown (Hurd *et al.*, 1986). The metaphor of a voyage is apt. During the *Age of Discovery* (i.e. 15th to 18th century), the Europeans not only explored the *New World* for markets, minerals, and colonies; they also hoped to discover new knowledge during their voyages and numerous scientific expeditions.

Curiosity prompted Charles Darwin (1859) to join the *HMS Beagle* to explore South America. He *collected* many plant and animal specimens along the way and *observed* that there were variations between, for example, tortoises on the Galapagos Islands and those on the mainland. Darwin spent many years *reflecting* on this variation. In the process, he posed that question of whether there are *limits* to animal or plant populations. He then proposed a *theory* of natural selection, a process where plants and animals *adapt* to changes in their environments. Those living organisms that adapt better will have more offspring and, over time, the better-adapted traits will replace old ones.

Critical Science

Science is critical in two senses. First, it is critical in the sense that it is open to public scrutiny of its techniques. Scientists use this feedback to improve the theory, and this enhances the quality of scientific work. The "public" here is often fellow scientists who understand the work and serve as reviewers for journals or speak out at scientific conferences. These diverse views also provide checks and balances against unfair criticism and scientific fraud.

Second, science is critical in the sense of *empowering* rather than enslaving people. This view of science has its origins in critical theory associated with the Frankfurt School (Marcuse, 1964; Habermas, 1971). For critical theorists, the task of social theory is not to seek knowledge for its own sake. It is to find ways to alter existing social structures and conditions to improve people's lives. Hence, critical theory is a critique of capitalist society. It does not take existing conditions for granted. It seeks ways to unmask certain cultural assumptions to improve our ways of life. For example, the teacher–student relation varies across societies. Critical theorists are in favor of tilting the balance of power towards students so that learning is creative, democratic, participatory, and inclusive (Anyon, 2008).

The Demarcation Problem

The demarcation problem in the philosophy of science concerns distinguishing science from non-science (or pseudo-science). The latter includes religion, myth, superstition, magic, art, literature, and other beliefs.

For the early Greek philosophers, what distinguished science from non-science is its focus on *causes*. This criterion is clearly insufficient to distinguish science from mere beliefs that may be causal. In the early 20th century, the Vienna Circle of philosophers argued that science deals with matters of fact and logic (Kraft, 1969). Scientific statements must be *verifiable*, that is, they are based on evidence.

Popper (2002) argued that no amount of evidence could verify or prove that a scientific statement is true. However, it is possible to falsify it, such as the observation of a black swan will falsify the statement "All swans are white." Hence, he proposed *falsifiability* as the criterion of demarcation. In other words, scientific theories are falsifiable, whereas non-scientific theories are non-falsifiable. For example, by Popper's criterion, religion is not a science because it deals with the spiritual. It is beyond the physical realm and is, therefore, not falsifiable by empirical evidence. Tribal magic is partly spiritual and partly based on ad hoc "cures" as evidence. If it does not work, there is something else for the witch doctor to blame, such as infidelity, to save the "magic" from refutation. A problem with Popper's demarcation criterion is that a non-scientific theory may be falsifiable, but only that it fails the test.

For Kuhn (1962), the demarcation criterion lies in *solving puzzles*. If a prediction fails, a scientist needs to solve the puzzle of why it fails. Hence, in addition to falsifiability, it is necessary to add the criterion of *progress* (Lakatos, 1978). A theory may progress through modifications that result in better interpretations, methodologies, explanations, or predictions.

Some philosophers such as Feyerabend (1975) and Laudan (1978) think that the demarcation problem is itself a pseudo-problem or, if it is not, it is still waiting for a solution. What is clear is that there is no single demarcation criterion.

What is Methodology?

A research methodology is a series of logical steps from formulating a research problem to arriving at a conclusion. It provides the link between theory and evidence, including the use of agreed standards to maintain rigor. Roughly speaking,

Methodology = Philosophy + Research designs + Methods.

The *philosophical* part of methodology implies that methodology is also a study of the research designs and methods used. It is not the designs and methods themselves, but an analysis of their use in a specific field of study.

Broadly, there are two main philosophies of science, namely,

- causal science; and
- interpretive science.

These philosophies underpin the research designs. Within each philosophy, there are several variants. For example, in causal science, there are positivist, post-positivist, realist, and conventionalist approaches. In interpretive science, there are interpretivism, hermeneutics, constr-uctivism, discourse analysis, grounded theory, critical theory, symbolic interactionism, ethnomethodology, gender, and phenomenological approaches. The differences within each philosophy of science are relatively minor compared to the difference between them. Hence, we shall stick to our main classification between causal and interpretive sciences.

The *research design* may consist of a

- survey;
- case study;
- experiment;
- regression; or
- comparison.

A research design is a systematic way of deciding how to execute the research to rule out alternative explanations. As discussed in the previous paragraph, it is underpinned by a particular philosophy of science. For example, an experimenter adopts a causal view of science to determine causes and effects. In contrast, an interpretive researcher using a case study design is interested in probing more deeply to gather different views on a topic.

The detailed *methods*, tools, or techniques are ways to collect and analyze the data. They are the nuts and bolts of research that occupy much of this book. Data may be collected using observations, interviews, ques-tionnaires, simulation, or past records. They may be analyzed qualitatively through techniques such as narratives or discourse analyses, or quantita-tively using statistics.

Causal Science

Causal science seeks to identify the connections between causes and effects, or the causal *mechanism* (Harre, 1970; Baskar, 1975). It shows how it works, or *explains* why it happens. This mechanism is often called a theory or hypothesis. A *theory* is a conception of how it works, including the assumptions and causal links that explain the event. A *hypothesis* is the testable part of a theory. For example, one may assume the absence of friction in physics, or that a consumer is maximizing his utility in economics. These assumptions are not tested; instead, what is investigated is the path of a projectile or the shape of the demand curve.

How do scientists explain? A *functional* explanation explains the existence of an entity or process by its function or effect; for example, slums exist because they function as a source of low-cost housing for poor workers. It is, however, not a causal explanation because there is no causal mechanism. A better explanation for the existence of slums that uses a causal mechanism will relate it to under-developed rural land, credit, and labor markets, or a collapse in agricultural prices, resulting in the rapid rural-urban migration of landless peasants. In the cities, the local governments are unable to cope with the provision of infrastructure for these migrants (Roberts, 1978).

In the search for causes, scientists distinguish between necessary and sufficient causes. For example, if a building catches fire, air is necessary but not sufficient to cause the fire. Further, they are seldom interested in just the *correlation* or regularity between any two events, such as the movements between interest rates and housing demand. Instead, they are interested in how changes in interest rates *affect* housing demand. In this case, a possible mechanism is that, all else equal, changing interest rates affect the cost of financing a house, and hence housing demand.

Unlike this example, many correlations are accidental and do not exhibit causal relations. For example, the prices of butter and bananas may rise at a similar rate but they do not cause one another. The cause of the price rise is a *confounding* factor: inflation.

Notice that the condition "all else equal" (or *ceteris paribus*) is used in the argument. This is because other factors such as household income and household formation also affect housing demand. This is a common problem in science. Nature is complex. There are many factors, and the challenge for the scientist is to isolate the factor under consideration by keeping other factors unchanged.

Causal science underpins several research designs such as

- experiments,
- regressions,

- comparisons, and
- case studies.

In an *experiment*, the researcher tries to manipulate one or more variables while holding other extraneous factors constant. For example, if the effect of a drug on cholesterol level depends on a subject's age, lifestyle, and gender, the researcher may fix age and gender by selecting a sample comprising only males aged 40 to 50. These extraneous variables are not of interest to the experimenter, but their effects on the outcomes of the experiment must be taken into account. The easiest way to do it in an experiment is to fix the values of these variables. The social scientist is not so fortunate. In the housing example, he cannot fix interest rates, household income, or household formation in the laboratory. His "laboratory" is society, and most factors are beyond his control. However, not all is lost. A possible solution is to use *statistical control* through a widely used technique known as *regression* (see Chapter 8).

A *comparative design* seeks to uncover common causes by examining a small number of different cases. This is Mill's (1884) method of agreement. If the cause is absent, then the effect is also absent. This is Mill's method of difference. For example, if factors *A*, *B*, and C are found in successful cases, then they are common causes. In failure cases, these factors will be absent. Mill recommended that scientists should use different cases, that is, they should be dissimilar. If the common causes can explain the occurrence of different cases, then the argument is more persuasive. For example, if natural resource endowment is identified as a possible factor that causes economic development, the scientist should look for countries with varying natural resources. Finally, if there are too many factors and too few cases, it is not possible to identify common factors.

A *case study* probes in-depth into a unit (case) to trace a process or discover something new. It may be causal, such as historical case study to trace the development of a unit or system over time. A case study may be interpretive. The researcher probes in-depth to understand the *context* and the actor's point of view. Here, context refers to the local situation or local knowledge or, more broadly, other factors. For example, Western management theories may not be applicable to firms in developing countries because of differing contexts where the politics, markets, and social norms are different.

In summary, the term "causality" can be confusing for the beginner. There are many ways of showing causality, and none is watertight. The weakest form of demonstrating causality is the use of *correlation* or *regularity*. This strategy is used in regression, experiment, and comparative research designs. Another way of demonstrating causality is to use causal *mechanisms*, which are used in case study research designs. Unfortunately, there may be competing mechanisms that explain the same event.

Interpretive science

The main differences between causal and interpretive science are shown in Table 1.1. For interpretivists, the world is not an objective external reality "out there", outside your head, to be discovered. Instead, the individual subjectively experiences and understands reality, resulting in *multiple realities* or different views of the same event (Collingwood, 1946; Taylor, 1971). These subjective understandings form the basis of human action.

Interpretivists seek to *discover* something new rather than some underlying mechanism, and often in a qualitative way. For example, they may wish to uncover a different perspective on a certain issue. Hence, unlike a causal scientist, an interpretivist does not usually test hypotheses. Instead, he uses an exploratory *framework* to discover new hypotheses, perspectives, and issues. A theoretical framework, or simply a framework, is an organized set of ideas. For instance, if we are studying the perspectives of

Table 1.1   Key differences between causal and interpretive science.

Feature	Causal science	Interpretive science
Reality	Objective; "Out there"	Subjective, in the actor's head
Purpose	Discover causal mechanisms	Discover something new
Strategy	Test hypothesis	Use exploratory framework
Design	Experiment, regression, comparison, case study	Case study, survey
Data	Numeric	Linguistic, symbolic
Data analysis	Statistics	Understand meanings

unskilled foreign construction workers on safety issues, a framework may consist of perceptions of risk, cultural differences, language barriers, safety training, sufficiency of experience, working hours, and so on.

The framework is then used to design the research, which typically consists of in-depth case studies of a few workers or broad surveys to gather different views. The linguistic or symbolic data are then interpreted to understand meanings or undercover the *reasons* the actors may have for doing something. These reasons consist of beliefs, preferences, attitudes, and values. It is also important to understand the local contexts in which people make decisions. For example, a local family business may not hire external managers for various reasons such as the desire to retain control and the general lack of trust in the local community or inadequate social capital (Putnam, 2001).

The interpretivist should be sensitive that each party will tell the story in a particular way, that is, from *his* point of view. Hence, the actor may deliberately *construct* the narrative through a careful choice of words (discourse) to persuade the listener. For example, the rhetoric (literary devices) of economics consists primarily of metaphors, analogies, and appeals to experts or data (McCloskey, 1986). Of course, any narrative may be *contested* or resisted by another party (Willis, 1977). A simple example is a dispute between the project owner and contractor. Both sides have their stories to tell.

In general, qualitative researchers are suspicious of numbers. They understand the importance of numbers, and do use numbers themselves. However, they are wary of the possible misuse of numbers by quantitative researchers. For qualitative researchers, what lies behind the numbers is a story or different stories as perceived or told, for one reason or another, by different actors.

The Research Process

The research process consists of the following steps:

- identify the *research question* (or problem);
- review the literature to develop a *hypothesis* in causal studies or *framework* in interpretive studies;

- determine an appropriate research *design* to test the hypothesis in causal studies or discover something new in interpretive studies;
- devise appropriate *methods* to collect data;
- collect and process data;
- analyze the data; and
- conclude and publish.

This process underpins the structure of this book. For this reason, we will discuss these steps in subsequent chapters.

Testing Theories

The test of a hypothesis is not a straightforward one. As Popper (2002) argued, the verificationists erred because no amount of white swans would prove the claim all swans are white. However, it takes only a black swan to refute it.

Unfortunately, things are not so simple. If we consider the issue whether governments or markets cause education, health, economic, political or social problems, opinions are largely divided between two camps, namely, conservatives who are in favor of free markets, and liberals who support greater government interference. There is, of course, a third camp of Marxists who want to replace capitalism altogether. For now, we shall restrict the discussion to the conservatives and liberals. There is no conclusive evidence that supports one camp or the other. For example, in the case of education, conservatives in a country may argue that the public school system is "failing" and education should be privatized. How do we know that the system is "failing"? Conservatives may argue that schools are inadequately funded, there is a shortage of good and qualified teachers, students did poorly in international tests and have discipline problems, and the curricula contain too little science and technology. The issue now shifts from testing whether governments have done too little or too much in education to testing each of these claims. Even if each claim can be tested, we may be left with a mixed bag where only *some* claims are supported by the evidence.

Now consider a particular claim, for example, whether the students did poorly in international tests such as the Program for International Assessment (PISA) test on reading, mathematics, and science for

15-year-old pupils. Here one may question whether the testing is objective, whether the tests that are based only on right or wrong answers are appropriate, whether other countries teach to the test (that is, they prepare their students for it), 'whether the questions take into consideration the local context, and whether more complex skills are being tested. In short, the testing is controversial because of disagreements over appropriate *measures* of competency in English, mathematics, and science. This is the correspondence issue, that is, whether a measure corresponds to the concept.

A related problem with measures is that some theoretical concepts (ideas) are not directly observable. In physics, we cannot directly observe absolute zero temperature, or zero K (Kelvin), that is, −273.15°C. The absolute pressure of gas is zero at that temperature. Unfortunately, gases liquefy and solidify long before reaching that temperature.

As a second example of the issue of unobservability, consider the problem of assessing creative student writing. In general, there are at least three ways of assessing creativity (Fleith *et al.*, 2014), namely,

- demonstration of divergent thinking;
- use of expert opinions; and
- use of criteria.

Measures of divergent thinking ask students to think of the many uses of an object, such as a safety pin. It is not appropriate for creative writing. The use of expert opinion to rate creative products is subjective. Hence, when teachers mark essays for evidence of creative writing, they use certain grading criteria. When I was in school, I was getting no closer to what constitutes creative thinking when my teachers spoke of originality, novelty, and ability to move the reader. Perhaps they were right, because the essays I had written were not memorable. I did not use tropes to persuade. Neither did I develop the characters or my story properly. Finally, I also did not appeal to the senses. There were no details.

In summary, it is difficult to settle debates in science based on a single theory or a single refutation. Other relevant claims (theories) may be involved, and we often end up with a mixed bag where there is only some empirical support. If the evidence is inconsistent with the theory, it is even possible to reject the evidence as flawed or adjust the theory to accommodate the contrary evidence. Hence, debates in science often rage on for many years until a theory increasingly does not fit the new evidence, in which case it loses more and more supporters, or a new theory that better-fits the evidence takes its place. Kuhn (1962) called it a paradigm shift or scientific revolution, where a "paradigm" is a group of related theories or a "school of thought." In our example, there is a clash between two schools, conservatives and liberals, over education policy.

From a different perspective, McCloskey (1986) has suggested that scientists *persuade*, rather than test, theories. If the falsifiability criterion is problematic, then scientists use a battery of tests and techniques to convince their fellow scientists that, on balance, their theories fit the data better.

Research as a General Competency

Some students take the view that knowing how to do research has only one purpose, and that is to write a dissertation or thesis required for graduation.

This view on the purpose of learning research methods is too narrow. If firms compete using technology as a core competence (Hamel and Prahalad, 1994), then employees who lack the generic competences to find causes, design a proper testing procedure, collect evidence, and draw valid conclusions are handicapped in acquiring new technology. They cannot make sense of the literature or evaluate the technology. It is also difficult to critically reflect on something or find creative solutions if one is ill-prepared.

For these reasons, many firms conduct general competency tests to assess an individual's cognitive ability to use reasoning skills to solve problems. These skills lie at the heart of research methods.

Many training institutes also rank research skills highly. For example, the College of Medicine at Florida Atlantic University, USA requires graduates to have the following general competencies:

1. Medical knowledge and research skills;
2. Patient-centered care;
3. Ethics and Law;
4. Professionalism;
5. Interpersonal and communication skills;

6. Cultural competency;
7. Health promotion and disease prevention for patients and populations;
8. Life-long learning and self-improvement;
9. Systems of health care practices;
10. Self-awareness and personal development; and
11. Community engagement, service, and advocacy.

Graduates are expected to contribute to medical knowledge. They should be able to identify knowledge gaps and seek appropriate solutions.

References

Anyon, J. (2008) *Theory and educational research: Toward critical social explanation*. London: Routledge.

Bhaskar, R. (1975) *A realist theory of science*. London: Reed Books.

Collingwood, R. (1946) *The idea of history*. London: Oxford University Press.

Darwin, C. (1859) *The origin of species*. London: John Murray.

Feyerabend, P. (1975) *Against method*. London: Verso.

Fleith, D., Bruno-Faria, M. and Alencar, E. (2014) *Theory and measurement of creativity*. Waco, Texas: Prufrock Press.

Hamel, G. and Prahalad, C. (1994) *Competing for the future*. Boston: Harvard University Press.

Habermas, J. (1971) *Knowledge and human interests*. New York: Beacon Press.

Harre, R. (1970) *The principles of scientific thinking*. London: MacMillan.

Hurd, D., Johnson, S., Matthias, G. and Snyder, E. (1986) *General Science: Voyage of discovery*. New Jersey: Prentice-Hall.

Kuhn, T. (1962) *The structure of scientific revolutions*. Chicago: University of Chicago Press.

Lakatos, I. (1978) *The methodology of scientific research programs*. London: Cambridge University Press.

Laudan, L. (1978) *Progress and its problems: Towards a theory of scientific growth*. Los Angeles: University of California Press.

Kraft, V. (1969) *The Vienna Circle: The origin of Neo-positivism*. Westport: Greenwood Press.

Marcuse, H. (1964) *One-dimensional man*. New York: Beacon Press.

McCloskey, D. (1986) *The rhetoric of economics*. Wisconsin: University of Wisconsin Press.

Mill, J. (1884) *A system of logic*. London: Longman.
Popper, K. (2002) *Conjectures and refutations*. London: Routledge.
Putnam, R. (2001) *Bowling alone*. New York: Simon and Schuster.
Roberts, B. (1978) *Cities of peasants*. London: Edward Arnold.
Taylor, C. (1971) Interpretation and the sciences of man. *Review of Metaphysics*, **25**, 3–51.
Willis, P. (1977) *Learning to labor*. New York: Columbia University Press.

CHAPTER 2

The Research Problem

Introduction

Research starts with a problem of interest to the researcher and to the community because of its scientific, social or economic value.

The research problem or question is often stated in the first chapter of the research report or the first section of a research paper. It comprises:

- a statement of the problem;
- its justification;
- the scope; and
- the objectives.

These components are discussed below.

Statement of Research Problem

The research problem is an issue or concern that warrants attention. It should be well articulated and stated as concisely as possible. The formulation of the research problem is the most important step in the research process and probably the most difficult as well. This is because many novice researchers are unclear as to what constitutes a well-articulated research problem. If the problem is unclear, the research loses its focus.

Normally, there is only one research problem, although it may be broken down into a small set of sub-problems. In most cases, it suffices to state the research problem and not the sub-problems. For example, Smith and Dejoy (2012) wanted to study how psychological and organizational factors affect workers' experience of safety and health issues in their work environment. It is a matter of preference whether to state the research problem or the research question. For this example, the research question is:

"How do psychological and organizational factors affect workers' experience of safety and health in their work environment?"

Similarly, Liebow (1967) wanted to know how poor, urban, adult, Afro-American males view themselves and the world around them, and how they adapt to it. In this case, it is possible to split the research problems into two parts: (1) on how they view themselves and the world around them, and (2) how they adapt to it.

Once the research problem has been stated, it is possible to state the tentative *title* of the study. For Smith and Dejoy, it is:

Occupational injury in America: An analysis of risk factors using data from the General Social Survey.

The title of Liebow's study is concise:

Tally's corner.

It refers to the street corner frequented by the poor men in Washington DC, with Tally as an important informant in the study.

State the title succinctly. You should read it repeatedly to remove unnecessary words. Finally, note that the title of the study is not the same as the research problem or research question in both examples.

Justification

The justification for the research answers the question "why are you studying this?" There should be clear justification on why a researcher is carrying out a particular research. In other words, the study should make a significant impact to the community. It may be based on theoretical or practical grounds, such as by contributing to knowledge or solving practical problems. It should not be of a personal nature. For example, doing research to earn a degree or learn about a particular topic does not constitute a justification for research.

For our two examples, Smith and Dejoy (2012) argued that occupational safety and health remain a significant problem because of the large number of injuries and the consequent economic costs (Weil, 2001;

Schulte, 2005). Moreover, most studies of the distribution of occupation injury use various job and employment factors (Dembe *et al.*, 2004; Simpson *et al.*, 2005). Much less is known about other risk factors, particularly those pertaining to psychological and organizational factors.

Likewise, Liebow (1967) provided three justifications for his study. First, the problems faced by and generated by low-income urban families are of major concern to policy makers in America (Orshansky, 1965). These concerns are nothing new, and existed for many years. Charities have been trying hard to help resolve them, but they are unable to deal with the scale and complexity of the problems. Second, much of what we know about poor Afro-American families is biased towards women and children, with a corresponding neglect of adult men. This is because the men are harder to reach than women and children (Moynihan, 1965). Third, much of what we know about the problems were gathered through surveys using interviews and questionnaires, which tend to provide only superficial understandings of the problem (Rohrer and Edmonson, 1960). For Liebow, what was needed was an in-depth probe through participant observation, of how these men view themselves and the world around them in their own terms, and how they adapt to it.

These two examples share some common features. First, the researchers are of the view that what they are trying to solve is a significant problem, namely, workplace safety and health, and urban poverty. Second, in justifying their research, they point to the research gap by briefly identifying the relevant literature. In the case of Smith and Dejoy (2012), current studies do not take into account the psychological and organization factors. For Liebow (1967), the research gap consists of two issues: they are the neglect of urban adult Afro-American men in previous poverty studies, and the lack of in-depth probes into the perspectives of these men.

How do these researchers know the research gaps? They may arise from personal or work experience, informal discussions with experts, casual observations of data patterns, a desire to challenge folklore, reading the news on a new process, product, or policy, intuition or, more commonly, a brief survey of the literature. The researcher reviews what is known about the problem and then tries to figure out the research gaps. In the process, he may have some hunches or intuition about the issue, and stands to gain substantially by consulting seasoned researchers on how to go about crafting the research question.

For the beginner, this type of feedback is important. Sometimes a simple remark that "this problem is not researchable" or that "it has been well researched" will set you thinking critically instead of blindly searching for a research problem. More often, the experienced researcher will hint to the student to refine his research question. This may take many iterations, and one must take it positively as a learning process.

Scope

The research problem does not fall from the sky. It usually starts with a broad research topic. The researcher then scopes it into manageable terms so that the problem is solvable within time and cost constraints.

For Smith and Dejoy (2012), the scope is clearly restricted to how psychological and organization factors affect workers' experience of work injury. They deliberately do not wish to consider other factors either because we already know a lot about their effects or because of resource constraints. Another possible way to scope the study is to restrict it to workers in a particular industry.

Liebow (1967) delimited the scope of the research to poor Afro-American men who occupied a street corner in the Second Precinct of Washington DC. He is clearly not studying all types of poor men from different backgrounds. He restricted the study to a particular race. This raises the issue of whether the study is generalizable, given that (1) it is an American setting in the early 1960s, (2) only poor Afro-American men were involved, and (3) he chose only a precinct of Washington DC. Further, Liebow did not consciously select the men for representativeness because he did not intend to develop generalizations. His research strategy was to probe intensively into the lives of these men. This argument, as we have learned in Chapter 1, is a classic defense of interpretive small-N case studies.

Objectives

The research objectives indicate what the researcher expects to *achieve* by doing the study. For Smith and Dejoy, the objectives are to:

- examine the impact of psychological and organizational factors on workers' experience of work injury; and

- to recommend ways to reduce the rate of injuries.

  In Liebow's case, the objectives are to:

- explore in-depth on how the street corner men view themselves and the world around them, and how they adapt to it; and
- recommend ways to help these men escape poverty.

Usually, there are two or three research objectives. We often want to determine, test, assess, analyze, develop, identify, estimate, compare, or ascertain something. The research process is not a research objective. For example, "to review the literature" is not a research objective. The review is what you need to *do*, not what you intend to *achieve*.

Organization of Study

The final section of the first chapter of a research report contains a short description of how you have organized the study. For instance, you may state that:

> *Chapter 2 provides the literature review and develops the hypothesis. The methodology is given in Chapter 3, where it consists of a regression model using a sample of 200 houses. Data on house prices and characteristics were collected using valuation reports. Chapter 4 provides the regression results, and Chapter 5 concludes the study.*

This section should be brief, as the details will be discussed in each chapter. It merely provides a road map of what is in the research report.

References

Dembe, A., Erickson, J. and Delbos, R. (2004) Predictors of work-related injuries and illnesses: National survey findings. *Journal of Occupational and Environmental Hygiene*, **1**(8), 542–550.

Liebow, E. (1967) *Tally's corner*. Boston: Little, Brown & Co.

Moynihan, D. (1965) *The Negro family: The case for national action*. Washington DC: US Department of Labor.

Orshansky, M. (1965) Counting the poor: Another look at the poverty profile. *Social Security Bulletin*, January, 3–29.

Rohrer, J. and Edmonson, M. (1960) *The eighth generation: Cultures and personalities of New Orleans Negroes*. New York: Harper and Row.

Schulte, P. (2005) Characterizing the burden of occupational injury and disease. *Journal of Occupational and Environmental Medicine*, **47**(6), 607–622.

Simpson, S., Wadsworth, E., Moss, S. and Smith, A. (2005) Minor injuries, cognitive failures and accidents at work: Incidence and associated features. *Occupational Medicine*, **55**(2), 99–108.

Smith, T. and Dejoy, D. (2012) Occupational injury in America: An analysis of risk factors using data from the General Social Survey (GSS). *Journal of Safety Research*, **43**(1), 67–74.

Weil, D. (2001) valuing the economic consequences of work injury and illness: A comparison of methods and findings. *American Journal of Industrial Medicine*, **40**(4), 418–437.

CHAPTER 3

Framework or Hypothesis

Literature Review

Recall that a theoretical framework, or simply a framework, is used to guide an interpretive study while a hypothesis is tested in a causal study. In both cases, the researcher needs to review the literature to:

- develop the framework or hypothesis;
- discover productive ways to improve the methodology and data analysis; and
- familiarize himself with previous findings.

There are two stages in a literature review. The first stage is the *preliminary review*. The aim is to identify the research gap and justification for the study. We have discussed this part of the literature review in Chapter 2 of this book.

The second stage is the *detailed review* where the purpose is to *develop* the framework or hypothesis. Here, the literature review should not "cover the field" or produce a "shopping list" of works by different authors. A long literature review is unnecessary, particularly if it contains peripheral works. It suffices to review just the key concepts to develop the framework or causal mechanism.

The literature review is also an opportunity to discover productive ways of doing your research so that you can benefit from previous studies. For example, if you are testing a hypothesis, you should review how the concepts are measured (operationalized), how sampling is done, the sources of data, and how they are processed and analyzed.

The final purpose of the literature review is to familiarize yourself with previous findings so that you can compare your results. It is important to pay attention to the local contexts so that the comparison makes sense.

It is good practice to summarize each book or journal paper on a card or a piece of A4-size paper. You can then arrange the cards and sheets alphabetically to form the bibliography. It is more efficient to review the latest publications first rather than start from older publications. The more recent publications would have reviewed earlier studies.

Framework

I will use Liebow's (1967) study as the first example of how to develop a framework for the study. Recall from Chapter 2 that he wanted to understand how poor urban Afro-American men view themselves and the world around them, and how they adapt to it.

Which framework should Liebow use? He used the *roles* of these men in their daily lives as workers, fathers, husbands, lovers, and friends to organize the study. These roles are structural positions in society, and they come with certain responsibilities and expectations. Sometimes, roles may not be clear-cut. For instance, what is the role of government in society? Here, opinions differ among conservatives who prefer limited government, to liberals and radicals who support greater government intervention either to correct market failure or replace it with central planning (Clark, 2016).

When the men play their roles, they need to adhere to certain behavioral *norms* or social rules. In turn, these norms are derived from community *values*. For example, if a community values the family, then fathers have certain roles to play and should adhere to generally agreed rules of behavior.

The combination of society's values and norms is what we call *culture*. This is a narrow but useful definition of culture. A broader definition of culture is a way of life that includes a society's political, economic, and social systems, food, art, clothes, and so on. Such a broad definition is not useful from a research viewpoint because it lacks focus. For example, it is difficult to measure culture if it is defined so broadly.

In any society, different groups (for example, social classes) will tend to exhibit different cultures. Thus, we speak of Malay culture, working class culture, bourgeois culture, corporate culture, and so on. These groups may agree on *core values*, but each group may still hold other values. They then *socialize* their children into these values, that is, children *learn* to behave in ways expected of them.

If these street corner men need to play certain roles, what happens if they are unable to fulfil them? They may feel frustrated or even ashamed, and may walk out of the family altogether. They may have a poor *view of themselves*, and think society holds a similar opinion. But people do want to improve their lot, especially that of their children. What, then, are the *structural constraints* that prevent poor people and their children from making it in society?

In summary, role theory is the framework, but this alone is insufficient. The researcher needs to develop the framework by enriching it with related ideas. These ideas are the ones in italics above. Hence, Liebow needs to establish the values, rules, expectations, perspectives, and constraints in the different roles.

Our second example on how to develop a framework to guide the research concerns the study on smart cities. The idea is relatively new, and Singapore is at the forefront of developing a smart nation under its recent Smart Nation Program (2014). Here, we are not looking for causes of smart cities. Rather, the study is exploratory, and the goal is to *understand* and *explore* the potential of smart cities in using informatics and technology to improve the organization and efficiency of urban services. Consequently, a framework is necessary to guide the exploration of what is a smart city and how it works. It may include the definitions of smart cities, types of smart cities, goals, defining features, development of a smart city ecosystem of firms and workers, pilot programs, integrative standards, costs, equity, inclusiveness, and other issues (McLaren and Agyeman, 2015; Stimmel, 2015).

We then probe each element of the framework further. For example, under "other issues," it is possible to consider privacy issues arising from the use of sensors throughout the city. This is particularly so if multiple data streams are continuously captured and analyzed by a single government agency. There are also security issues. If the online Smart Nation Platform is hacked, then hackers or criminals may have access to large amounts of data. Similarly, there will be many pilot programs to test the smart city concept, such as telemedicine, smart homes, smart grid, smart car parking, smart mobility, and smart street lighting.

Hypothesis

In the case of a hypothesis, the literature review is used to develop the causal mechanism. It may be in textual, symbolic, mathematical, or physical form.

As an example, consider a causal study of the housing bubble in a particular city over a certain period. It is clear what is at stake in a housing bubble. The bubble affects the economy by increasing household wealth, which in turn increases consumption of other goods and services. However, it may erode the work ethic because of potential profitable speculation. In addition, the collapse of a housing bubble may put the financial system in danger because of excessive bank lending to the housing sector.

However, it is less clear what constitutes a housing bubble. Thus, the literature review will need to start by defining a housing bubble. The issue is not just definitional or academic. The practical implication is that housing experts may disagree whether a housing market has a bubble in the first place. If there is no bubble, there is no basis for any government policy to correct "inflated" prices.

The usual understanding of a housing bubble is a rapid rise in house price, usually relative to benchmarks such as:

- household income;
- rents;
- user cost; or
- its fundamental value.

If disposable household income is stable and house prices rise rapidly over the same period, there is evidence of excessive price appreciation. More commonly, both house prices and household incomes rise over the same period. Hence, we need to compare the ratio of median house price (P) to median disposable household income (Y). Usually, we use the median rather than the mean value to eliminate the effects of outliers. The median is the middle score, and it is unaffected by extreme values. In Figure 3.1, we see that the ratio P/Y varies between 6 and 15. The question is this: beyond which level in P/Y do we consider house prices to be in bubble territory? Here, opinions differ. According to the

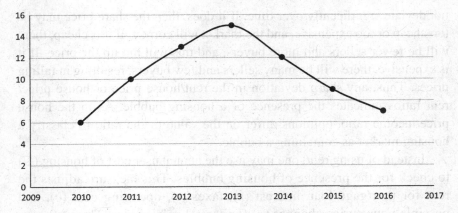

Fig. 3.1   Variation of house price/income ratio, 2010–2016.

Demographia International Housing Affordability Survey (Demographia, 2017), the results are as follows:

US	3.6
Canada	3.9
Japan	4.1
Singapore	4.8
Australia	5.5
New Zealand	5.9
Hong Kong	18.1

Demographia defines a ratio greater than 5.1 as "severely unaffordable." By this measure, Australia, New Zealand, and Hong Kong fall into this category. The ratio for Hong Kong is clearly an outlier. A related *P/Y* ratio considers only potential first-time buyers, which is more reflective of housing affordability.

Similarly, by viewing housing purchase as an investment, the house price/rent ratio or its inverse, the rent/house price ratio, should remain relatively constant. The rent/house price ratio is also called the rental yield, and it should not move too far out of line as a market rate of return on the investment. This is similar to investing in the stock market, where the dividend/price ratio of a company's share in a mature industry should not deviate significantly over time. If it does, then the share price may be too cheap or too expensive, and the market will correct. If it is cheap, there will be fewer sellers and more buyers, and this will bid up the price. If it is expensive, there will be many sellers and few buyers, resulting in falling prices. Thus, any sharp deviation in the rent/house price or house price/rent ratios indicates the presence of a housing bubble. As in the house price/income ratio, opinions differ on the value of the ratio to classify a housing market as containing a bubble.

Instead of using rent, one may use the annual user cost of housing (U) to check for the presence of housing bubbles. This measure adjusts the rent for mortgage loan interest (r), taxes (t), operating cost (c), and possible house price changes (g) for capital gain or loss. Thus, we may write:

$$U = (r + t + c - g)P$$

where the components are expressed as a percentage of house price (P).

Lastly, one may discount the annual future net rents a house can earn (R_t) and compute its fundamental value (V), that is,

$$V = R_1/(1 + k) + R_2/(1 + k)^2 + \ldots + R_n/(1 + k)^n.$$

The net rent is gross rent less outgoings. Here k is the discount rate, and n is the remaining lease of the house in years. Any large deviation between house prices and fundamental values indicates the presence of a bubble (Garber, 2000). There are two major difficulties in determining V, namely, the choice of the discount rate and problems with predicting future rental streams. A minority of researchers deny the possibility of bubbles; if prices rise substantially, even in a short period, then it is because the fundamentals have changed, not because of the presence of a bubble. The basis for this belief is the efficient market hypothesis (Malkiel, 2016). According to this view, there is no such thing as an over-valued or undervalued asset.

The next step in the literature review is to consider current explanations of housing bubbles. One group of explanation focuses on housing market imperfections, particularly supply side constraints that prevent the short-run housing supply from adjusting rapidly to rising demand during a housing boom (Maisel, 1963; Quigley, 1997). These constraints include difficulties in obtaining land, securing financing, collecting housing market information, and hiring of construction workers in a boom market.

Another group of researchers focuses on the demand side, either in terms of:

- fundamental factors such as demography, household income, and interest rates (Poterba, 1991; Hubbard and Mayer, 2009);
- psychological factors in terms of buyers' irrational expectations of future house prices (Case and Shiller, 2003); or
- the impact of purchases by out-of-town or foreign speculative buyers (Chinco and Mayer, 2014).

Finally, how does one test for the existence of housing bubbles? For researchers who stress fundamental factors such as demand and supply factors, a logical choice is to *regress* house price on these factors to determine their explanatory power (Kaufmann and Muhleisen, 2003). If the regression fit is low, then fundamental factors do not explain much of the house price change, that is, there is evidence of a bubble.

Another possibility is to test for structural change in the housing market, that is, we use two regressions, one before the rapid rise in house prices, and one during the boom period (Jones and Leishman, 2003). If there is a bubble, the regression coefficients will differ significantly between the two periods.

A third possibility is to test whether house prices move in tandem with fundamental factors or there are sharp deviations. The technical term is the asset pricing approach using co-integration analysis (Meese and Wallace, 1994; Arshanapalli and Nelson, 2008).

Lastly, for researchers who stress the importance of psychological factors, the preferred research design is a *survey* to ask recent house buyers of their opinions on whether a housing bubble exists (Case and Shiller, 2003).

Obviously, if you are doing the literature review, you have to decide which test to use by reviewing the strengths and weaknesses of each approach before making the decision. In addition, you will need to improve the model or apply it in a different context (for example, a different housing market) as part of your research contribution. The improved model is your research hypothesis. For example, if you decide to use a survey, you may wish to apply the Case–Shiller methodology to the housing market in your city rather than in American cities. Further, you should carefully review how the Case–Shiller survey was done and how it can be improved in your study. You can improve the hypothesis on how expectations of house prices are formed by first studying criticisms of the Case–Shiller methodology (Quigley, 2003).

In summary, the literature review for causal analysis is not an *ad hoc* list of authors or an attempt to cover everything on the topic. It focuses on the research problem, and on how to develop a hypothesis for testing. Only the key research publications or representative papers in competing explanations are cited to provide readers with a sense of how the main ideas are connected. The review may be organized as follows:

Definition of housing bubble

Explanations of housing bubbles

Supply factors

Demand factors

Fundamental factors

Psychological factors

Methods of testing for bubbles

Regression on fundamental factors

Tests for structural change

Asset pricing approach

Survey of buyers' price expectations

Hypothesis

Dependent variable

Independent variables

Recall that, other than developing the hypothesis, you should review the literature to improve the research methodology. Finally, you should also review previous findings so that you can compare your results.

Research Proposal

A research proposal is required to secure funding or approval. It normally consists of the following:

- Title;
- Research problem;
- Justification;
- Preliminary literature review;
- Research design;
- Data collection method;
- Data collection plan;
- Data processing plan;
- Data analysis plan;
- Schedule of costs;
- Schedule of deadlines (see Table 3.1);
- Risk assessment;
- Background of researcher(s) involved; and
- Bibliography.

The schedule of costs and background of researcher(s) are not required in student research proposals. These proposals are used to facilitate the allocation of thesis or dissertation supervisors.

Many research evaluators pay close attention to costs that may include the salaries of researchers and assistants, stationery, postage, photocopying, computing, equipment, materials, transport, books, and so on. It is

Table 3.1   Example of research schedule.

Tasks	Months					
	2	4	6	8	10	12
Problem formulation	▓					
Literature review & hypothesis		▓				
Design & method			▓			
Data collection & processing				▓	▓	
Data analysis				▓		
Writing of research report						▓

important to cater for contingencies because research is a risky endeavor. For instance, the response rate for the survey may be too low, and this may necessitate a change in research strategy, such as by interviewing a new group of respondents.

Research Ethics and Risk Assessment

The researcher or a competent external party is normally required to conduct a risk assessment of possible harm or discomfort to humans and ways to manage these risks. A formal assessment requires the computation of the risk/benefit ratio. Individuals can then decide whether they wish to participate in the project, through informed consent in the sense that they are aware of the research purpose, its procedures, risks, and benefits. There is also the right to withdraw from the study without giving any reason. Obviously, not everyone is capable of giving informed consent. For example, for studies on children or the elderly, the researcher will be required to seek the consent of their parents or caregivers and take special care to protect subjects from possible harm or discomfort.

The risk assessment should also cover possible damage to property and equipment such as through negligent use, accidents, or fire.

The following information, if any, should be provided to potential participants:

- expected time commitment;
- boring or repetitive tasks;
- considerable physical exertion;
- physical harm, such as exposure to dangerous chemicals;
- privacy issues, such as income or access to medical or employment records;
- confidentiality, such as a worker's "feedback" on management or that names will not be disclosed in the research publication;
- recall of distressing events;
- possible allergies;
- measures of performance, to avoid embarrassment to poor performers; and
- studies that may link performance or deficiencies to nationalities, race, ethnicity, religion, income, or gender.

The researcher should provide full disclosure. A debrief is also conducted soon after the study so that respondents can assess whether the objectives of the study have been achieved.

Researchers should not use deception to gather field data. These techniques include misrepresenting the purpose of research or the level of risks, concealed observation or secret recording of behavior. For example, should a researcher conceal his identity when studying the behaviors and activities of a street gang? Some researchers do not reveal their identities until after the study has been completed because they believe it will affect their membership into the gang. Sound judgment is required; for example, in mass observation studies of pedestrian behavior, concealed observation from an overhead bridge is usually not an issue. However, if you are observing how a project manager conducts a site meeting, you should reveal your identity and obtain informed consent.

Plagiarism is the stealing of other researcher's data or work and presenting them as one's own work. It is a serious offence, and you should give proper credit when it is due. This does not mean that one should cite everything, which is overkill. For example, it is well known that Marxism originates from Karl Marx.

Do not falsify or fabricate the data. A more common problem is the deliberate or inappropriate deletion of outliers to make the results look better, such as by obtaining a better regression fit. This is a pity, because outliers, if they are not obvious mistakes or mismeasurements, play a major role in scientific discoveries. They provide a possible refutation of the theory.

Many universities and research institutions have Institutional Review Boards to tackle such issues on research ethics. The principal investigator should ensure that research assistants are properly selected and trained, and that they follow protocols.

References

Arshanapalli, B. and Nelson, W. (2008) A cointegration test to verify the housing bubble. *International Journal of Business and Finance Research*, **2**(2), 35–43.

Case, K. and Shiller, R. (2003) Is there a bubble in the housing market? *Brookings Papers on Economic Activity*, **2**, 299–342.

Chinco, A. and Mayer, C. (2014) Misinformed speculators and mispricing in the housing market. *NBER Working Papers* 19817. Massachusetts: NBER.

Clark, B. (2016) *Political economy: A comparative approach.* New York: Praeger.

Demographia (2017). *The International Housing Affordability Survey.* St Louis: Demographia.

Garber, P. (2000) *Famous first bubbles: The fundamentals of early manias.* Cambridge: MIT Press.

Hubbard, R. and Mayer, C. (2009) The mortgage market meltdown and house prices. *B. E. Journal of Economic Analysis and Policy*, **9**(3), 1–45.

Jones, C. and Leishman, C. (2003) Structural change in a local housing market. *Environment and Planning A*, **35**(7), 1315–1326.

Kaufmann, M., and Muhleisen, M. (2003) Are house prices overvalued? *IMF Country Report No. 3/245.* Washington DC: IMF.

Liebow, E. (1967) *Tally's corner.* Boston: Little, Brown, and Co.

Maisel, S. (1963) A theory of fluctuations in residential construction starts. *American Economic Review*, **53**, 359–383.

Malkiel, B. (2016) *A random walk down Wall Street.* London: W. W. Norton.

McLaren, D. and Agyeman, J. (2015) *Sharing cities: A case for truly smart and sustainable cities.* Cambridge: MIT Press.

Meese, R. and Wallace, N. (1994) Testing the present value relation for housing prices: Should I leave my house in San Francisco? *Journal of Urban Economics*, **35**, 245–266.

Poterba, J. (1991) House price dynamics: The role of tax policy and demography. *Brookings Papers on Economic Activity*, **22**(2), 143–183.

Quigley, J. (1997) *The economics of housing.* Northampton: Edward Elgar.

Quigley, J. (2003) Comment on Case and Shiller: Is there a bubble in the housing market? *Brookings Papers on Economic Activity*, **34**(2), 343–362.

Stimmel, C. (2015) *Building smart cities.* London: CRC Press.

CHAPTER 4
Research Design I: Case Study

Features of Case Studies

A case study is usually an interpretive study that seeks in-depth or intensive investigation of a particular case to *discover* or explore something new rather than *test* a hypothesis.

A case study tells a big story through the lens of a small case (Walton, 1992). A "big" story is a significant one, and there is a "story" in every case study. Without a storyline, it will be difficult for the reader to figure out the main message. A story contains a plot, which implies the existence of a structure to tell it persuasively.

As a research design, case studies differ from *case methods* that are widely used in business schools to teach students. In case methods, students study a written business case in advance of each class and debate the issues in class. A case method is a form of teaching and learning as an alternative to lectures, role playing, and other methods of instruction. It is not a research design. The main goal of case methods is to engage students in their learning on a real or imaginary business case. It is not to do research to solve a scientific problem, which concerns us here.

A "case" is a unit of study. It may be a person, team, project, organization, province, country, process, activity, or situation. Examples of processes, activities, and situations include recruitment processes, class teaching, and project disputes. The unit should be a clearly defined and bounded system.

Many case studies are interpretive, where the actor's perspective guides the research. Some case studies, such as a historical case study of the development of a town or organization, are causal. They trace the process (mechanism) linking causes to outcomes. However, unlike experimental studies, there is little control over extraneous variables in a historical case study. This makes it vulnerable to alternative interpretations of history. The historian has no control over what has happened in the past. The lack of control and methodological rigor is a general criticism of case studies. Nonetheless, the use of a case study to probe intensively to discover new insights is a major strength.

To be able to probe in-depth, a case study requires a large amount of data. Qualitative researchers strive to provide "thick" descriptions of the local context to provide the reader with a sense of "being there" (Geertz, 1973) so that they can appreciate the complexity of the situation. Without a proper understanding the context, certain behaviors or activities may not be meaningful to an outsider. A thick description is also *holistic*, that is, the researcher puts the various parts together in a coherent manner. In contrast, a thin description is one where the explanation is superficial.

Researchers who conduct case studies may use both qualitative and quantitative data. Understandably, qualitative data are more common in interpretive case studies. In causal case studies, such as in tracing the development of an organization or city, qualitative data also predominate.

Many disciplines in the social sciences, such as sociology, political science, business management, anthropology, psychology, and education, regularly use case studies to study phenomena (Liebow, 1967; Merriam, 1998; Samuels, 2012).

Sampling

The researcher needs to justify why he chooses a particular case or a small number of cases. There are several options.

First, one may select a *typical case* such as how a village school in a particular region operates. Being typical, it allows a limited form of generalization about how village schools in the region operate, even though it is only a single case. The extent of generalization will depend on how "typical" is the selected case.

Second, the researcher may select a *unique case* such as a village school in the region to discover why it excels. The purpose is no longer to generalize the findings. It is to discover why it is unique.

Third, one may select a *test case* to illustrate a principle, explore new issues, suggest new variables, refute a theory, or develop new ways of looking at the same facts or issues. The test case may be an outlier, in which case it does not conform to existing theory.

Finally, *multiple cases* may be used in situations where the researcher:

- is unsure if the selected case it typical, in which case he may use another case;
- wants to compare and contrast different cases, such as the different ways in which village schools are run; and
- wishes to *sample theoretically* to build theory, to generate new ideas from a small number of cases until the point of diminishing returns (Glaser and Strauss, 1967). This is the point of theoretical saturation, where new cases generate few ideas.

Causal Case Study

As an example of a causal case study, I consider Stone's (1989) study of the urban regime in Atlanta in the United States from 1946 to the 1980s.

An urban regime is an informal but stable governing coalition comprising leading politicians (often the mayor), senior bureaucrats, business leaders, and possibly the labor union leaders and other interest groups. A regime may:

- support economic growth (developmental regime);
- maintain the status quo (maintenance regime); or
- protect the environment (progressive regime).

There is no formal structure, but the developmental regime that Stone considered was interested in the redevelopment of the city in the face of global competition for capital investment.

Here, the context matters. Federal urban policy began with the Housing Act of 1937 to eliminate unsafe and insanitary housing. After World War II, the Act was amended to include slum clearance and urban renewal, which resulted in the destruction of working class communities and subsequent building of dense and high-rise public housing blocks in remote places (Anderson, 1967). In addition, the large concentration of poor and mostly unemployed people in these housing estates was a potential time bomb.

The Johnson administration (1963–1969) declared its War on Poverty by implementing federally supported programs to help the poor, racial minorities, and elderly through training and education, food stamps, insurance, social security benefits, and better economic opportunities. There was a policy shift from building large-scale mass public housing towards smaller projects, the preservation of heritage, and conservation. Despite these efforts, many urban riots took place in American cities during the 1960s.

In a major shift in urban policy, the Reagan administration (1981–1989) no longer viewed the "city problem" as a federal problem. It then cut categorical grants sharply, leaving cities to fend for themselves (Stoesz, 1992). The other plank of Reagan's urban policy was deregulation and privatization, which encouraged local governments to collaborate with the private sector to revitalize ailing cities. An example of such collaboration was the establishment of urban enterprise zones. The local government would provide tax and other incentives to encourage firms to locate in poor areas.

Consequently, during the 1980s, there was a substantial shift in the fortunes of American cities. Cities in the old "rust belt," such as St. Louis, Detroit, Cleveland, and Pittsburgh, lost about 22–31% of their population between 1970 and 1984, while cities in the South and West, such as Los Angeles, Houston, and Phoenix gained between 10–46% over the same period (Stoesz, 1992).

For Stone, nothing just happens for these declining or growing American cities. Neither the free market nor the tide of globalism alone shapes the future of cities (Petersen, 1981). The visible hand of the urban regime is what makes urban redevelopment possible; conversely, it may also end in failure through poor leadership and management, corruption, and other reasons. Hence, the capacity of an urban regime to effectively organize and revitalize a city is what matters.

Stone subdivided the research question regarding the urban regime in Atlanta into three possible sub-questions:

- who formed the governing coalition;
- how did it accomplish its mission; and
- what were the consequences.

He answered these questions by examining the actors, structure, ideas, and processes in a causal way in terms of how they evolved historically. The narrative is an analysis of the long political struggles and conflicts before a bi-racial governing urban coalition that supported urban development emerged.

On the issue of sampling, why did Stone select Atlanta? He argued that Atlanta's urban regime exceled in getting strategically important people in the city to act together. Thus, he is using a unique case to discover how the urban regime was formed, and how it managed to act coherently.

Can we generalize the study of urban regimes beyond Atlanta? Critics argue that urban regimes may not exist outside the United States because of differing political contexts (Keating, 1991).

Interpretive Case Study

Our example of an interpretive case study is Peshkin's (1986) study of the "total world" of a fundamentalist Christian school, the Bethany Baptist Academy (BBA, a pseudonym), in Illinois, America. The school immersed itself in a "total world" with its church and Christian families, with little contact with outsiders. Other examples of these "total institutions" include nursing homes, mental hospitals, military training camps, and prisons (Goffman, 1961). The film *One Flew Over the Cuckoo's Nest* is based on Goffman's idea of total institutions in public mental hospitals.

The objectives of Peshkin's study were to discover what such a school was like, what made it attractive to many American families, and the strains and tensions within the school. The *framework* comprised:

- the school's doctrine of a single truth, as outlined by its founding principal (pastor) and implemented by its headmaster,
- the Christian teachers,
- the structure of control,
- its socializing regime,
- the students' beliefs,
- the impact of the Christian school on its graduates,
- the students who did not fit well,

- the school as a total institution, and
- the benefits and costs of such a system to American society.

Peshkin structured the chapters of the book based on the above ideas. Except for the last chapter on benefits and costs, he wrote the chapters from the actor's perspective. For example, in the chapter on the Christian teachers (Chapter 3), Peshkin first provided the background of the 12 full time and part-time teachers. They came from large middle-class families in different US states and backgrounds, and most were married. All of them attended college and believed that God called them to be teachers. It was a "calling." The typical week began on Sunday, which was replete with church-related activities. From Monday to Friday, other than the usual classes, the teachers performed various tasks such as collecting children before school and driving them home after school, driving the soccer team to and from competitions, community visitation to spread the Christian message, and afternoon sports duty. On Saturday, the men had prayer breakfast followed by bus visitation. Peshkin noted that the teachers engaged in these activities quite willingly because of their "calling."

What did the teachers do in their free time? Few teachers were serious readers, and only one belonged to any organization outside the school church. Why? Because "Church and church friends take all my time." (p. 71) An average teacher's schedule could be broken down to having about 65% of their waking time spent in school/school-related activities, 15% in church, and the remaining 20% with family or friends. Importantly, all the teachers were contented with the busy schedule. Nonetheless, there were a number of strains, such as not having privacy in one's dating life — but most of the teachers were happy with the friends they have in church and school.

On the issue of pedagogy, teachers mined the Bible for pedagogic principles, believing in its richness. As Headmaster McGraw put it, "the Word of God should permeate every moment of the school." (p. 75) For example, in the teaching of science, the goal was to develop a Christian mind so that students see everything from God's perspective. Teachers also viewed students as "clay," and were challenged to shape their character. They looked for signs that identified the unsaved, such as their inability to pray or the use of inappropriate language. Once the teacher has identified a student, he would use a plan of salvation to engage the student in a conversation.

Peshkin then went on to probe how teachers viewed their relations with other stakeholders. Students were seen as disciples. Parents are viewed as co-workers, colleagues as friends, and administrators as leaders. They were loyal to administrators, and avoided gripes that demoralized.

How did the teachers contrast their work with those in public schools? They acknowledged that public schools were better resourced, and that their teachers were better paid. However, there were clear disadvantages in teaching in public schools such as poor discipline, social problems, less dedicated teachers, indifferent parents, and the inability to develop a student's character and teach the truth. For these teachers, the Christian school was just about right, and they were happy teaching there.

I hope this brief summary of Chapter 3 of Peshkin's book has provided a glimpse of the richness of his study and the importance of understanding the local context, particularly of the link between the church and school in this case study. The two entities are inseparable. It also illustrates how the base idea of the Christian teacher in the framework is further developed into related concepts such as how the teachers saw the world around them, their goals as teachers, the activities they did on a weekly basis, their pedagogies, their relations with other stakeholders, and what motivated them to teach in a Christian school. While Peshkin did have a viewpoint, he let the actors speak for themselves.

On the issue of sampling, why did he choose BBA? Peshkin, a Jewish professor of comparative education, was fascinated with religious schools. He chose BBA because permission was granted after several rejections by other Christian schools. A refusing pastor likened Peshkin to "...a Russian who says he wants to attend meetings at the Pentagon — just to learn... No matter how good a person you are, you will misrepresent my school because you don't have the Holy Spirit in you." (p. 12)

This example raises a number of questions. It is beyond doubt that an intensive study of an American fundamentalist Christian school is fascinating, as there are few such intensive studies. However, is Peshkin's portray of BBA as a total institution that served only God's Will accurate, given the cautionary misrepresentational remarks by the pastor? Further, can the study be generalized to other fundamentalist Christian schools?

References

Anderson, M. (1967) *The federal bulldozer*. New York: McGraw-Hill.

Geertz, C. (1973) *The interpretation of cultures*. New York: Basic Books.

Glaser, B. and Strauss, A. (1967) *The discovery of grounded theory*. Chicago: Aldine.

Goffman, E. (1961) *Asylums*. New York: Doubleday.

Keating, M. (1991) *Comparative urban politics*. Aldershot: Edward Elgar.

Liebow, E. (1967) *Tally's corner*. Boston: Little, Brown and Co.

Merriam, S. (1998) *Qualitative research and case study applications in education*. New York: Jossey-Bass.

Peshkin, A. (1986) *God's choice: The total world of a fundamentalist Christian school*. Chicago: University of Chicago Press.

Petersen, P. (1981) *The limits of the city*. Chicago: University of Chicago Press.

Samuels, D. (2012) *Case studies in comparative politics*. London: Pearson.

Stoesz, D. (1992) The fall of the industrial city: The Reagan legacy for urban policy. *Journal of Sociology and Social Welfare*, **19**(1), 149–167.

Stone, C. (1989) *Regime politics*. Kansas: University Press of Kansas.

Walton, J. (1992) Making the theoretical case. In C. Ragin and H. Becker (Eds.), *What is a case?* (pp. 121–137). Cambridge: Cambridge University Press.

CHAPTER 5

Research Design II: Survey

Features of Surveys

A survey is a research *design* that may be used in descriptive, interpretive, or causal studies. In descriptive surveys, the intent is to infer population characteristics from a subset of the population, called a sample. This is the most common understanding of a survey because of its association with statistical sampling. Students learn of such surveys from introductory classes in statistics.

Researchers also use surveys for interpretive and causal purposes. They ask respondents for their viewpoints on a particular topic, preferences on a range of options, or reasons for their actions. For this reason, a survey is a type of research design. It is not a *method* of data collection. To collect survey data, we use measuring instruments such as questionnaires and observation devices.

Typically, a survey uses a large sample to gather broad and relatively shallow sample characteristics. This contrasts with the case studies discussed in Chapter 3 that use a small sample to probe deeply into a complex issue. In other words, a survey sacrifices depth for breadth.

Surveys can also generate quantitative variables such as income and education attainment. We often correlate these variables with respondents' opinions on the issues. For example, a researcher may correlate voting decisions with voter income, party affiliation, education level, gender, age, and so on.

It is also possible to survey animals, plants, equipment use, rainfall patterns, building conditions, and so on.

Surveys are popular because they provide a quick and efficient way to obtain broad answers based on a sample before generalizing it to the population. The weaknesses of surveys include possible researcher, sampling, and response biases. They are also less appropriate if in-depth answers are required.

Types of surveys

Surveys may be ad hoc or carried out at regular intervals. Most surveys are *cross-sectional studies* that gather information about a population using a sample at a point in time. For example, we may ask a group of firms at a point in time on their organizational leadership, marketing strategies, or human resource policy. Such surveys are ill-suited for under-standing changes over time.

In a *longitudinal study*, we collect data over time to monitor changes. There are three types of longitudinal studies, namely,

- trend studies using different samples;
- cohort studies using different samples from the same cohort; and
- panel studies using the same sample.

For example, in a trend study, we may sample different consumers every five years to track changes in online shopping behavior over time.

A cohort is a group of individuals who share a common characteristic such as year/period of birth or class. For example, in a cohort study, we can take different samples from the same birth cohort (age group), such as those in their 20s, 30s, 40s and 50s.

Finally, a panel study traces the development of the *same* sample over time, such as starting from a sample of shoppers in their 20s and studying them every five years to track their shopping behavior. Another common use of a panel study is to increase the sample size to improve the precision of our sample estimates. For example, a researcher may not be able to access many firms in an industry. With only a few firms and a large num-ber of variables, it will be difficult to carry out a proper regression analysis using cross-sectional data. One way of overcoming this problem is to use panel data, that is, data collected from these firms over time (Hsiao, 2003).

Longitudinal surveys tend to be more expensive than cross-sectional surveys because of the need to monitor samples over time. In addition, if the same sample is used, there is a risk of participants dropping out of the survey.

Sampling

Sampling is a key part of every research design and not just of surveys. We have seen in Chapter 3 that, for case studies, we need to justify why we select a particular case or the small number of cases. In surveys, sampling consists of identifying the

- sampling unit or element;
- target population;
- sampling frame (if any);
- sampling method; and
- sample size.

A *sampling unit* or unit of analysis is the smallest unit of observation that is of interest to the researcher in a sample. In consumer surveys, the sampling unit is usually the household as the basic decision-making unit in terms of consumption expenditure. In business surveys, the sampling unit may be organizations or employees. A common problem in surveys is confusion over the choice of sampling units. For example, a newspaper article may report that most firms do not support a particular government policy such as raising the minimum wage. The problem here is that "firms" are business entities and, unlike people, they do not have opinions. We need to know who in the firm were interviewed about the policy — were they senior management, middle managers, or workers? In this example, the sampling units may be senior managers rather than firms.

The *target population* is the aggregate of all sampling units. In consumer surveys, it is called the target audience, the people for whom the product or survey is targeted. Another common term for target population is the theoretical population. The "theoretical" part is important because it underscores that it is merely a concept or idea. It is not the "real" thing. This means that we need to define the target population theoretically. As a simple example, we may want to survey poor people in the community for their views on a particular issue. There are two key concepts that require proper definition here, namely, "poor people" and "community."

How do we define "poor" people? We know that they lack material assets or money. The issue is where to draw the poverty line. To complicate the issue, people may be poor in the absolute sense, or relatively poor, that is, they are considered "poor" only when compared to others in the community. There are then two possible measures of poverty. Absolute poverty is often measured using the poverty line. For example, a person is considered poor if he earns less than $x a day. The value of $x is not fixed because of inflation but if a person earns less than this amount in *any* country, he is considered poor.

To measure relative poverty, we need to compare a person's income to some relative standard, such as the median income in the community. For example, a person is living in relative poverty in a particular community if he earns less than some fixed proportion of the median income.

Next, we consider how we can define the term "community." A community is a group of people who share some characteristics. The simplest definition is to use *place*, such as a village, town, or city. However, people living in the same city may have little in common. Hence, another way of defining community is, for example, by defining the group's way of life (Wirth, 1938). Finally, we can also define a community as a *network* of people sharing a common interest, belief, or cause, such as a community of scientists or stamp collectors. We can also find a virtual community of cyberspace gamers.

The *sampling frame* is the *list* of elements from which sampling takes place (Fig. 5.1). Unlike the theoretical population, which is conceptual, a sampling frame is the "real" thing, that is, it is an actual list. For example, the list of contractors from the local contractors' association is a possible sampling frame. It differs slightly from the theoretical population because not all contractors, particularly the smaller ones or foreign contractors operating in the city, belong to the association.

Many surveys do not have sampling frames. For instance, it is not possible to obtain a list of every fish in a lake. Similarly, if one is sampling

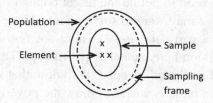

Fig. 5.1   Basic sampling concepts.

students in a large university, the sampling frame may consist of 30,000 names. Such a list is usually not available to the researcher. For this reason, we often use non-probability rather than probability samples. The rationale is discussed below.

As shown in Figure 5.1, the sampling frame should be as close to the target population as possible to reduce sampling bias. For example, if I wish to sample students from a particular university, a sampling frame comprising only students living on campus is not representative of the student population. Similarly, a list of undergraduates will not contain the names of graduate students.

Finally, it is important to use an updated sampling frame. For example, a list of firms in a dynamic industry may be quickly outdated. Updating can be resource-intensive, which explains why you as the researcher should check that you have an appropriate and updated sampling frame.

Probability Samples

Probability samples use random sampling to draw the samples from a sampling frame. As long as the sampling frame is close to the target population, probability samples are more accurate than non-probability samples. However, they tend to be costlier, such as the requirement of a sampling frame.

The main types of probability samples are:

- simple random samples;
- systematic samples;
- stratified samples; and
- cluster samples.

Simple random sample

In a simple random sample, we draw the sampling units randomly. Despite its simplicity and wide discussion in elementary statistical textbooks, we rarely use simple random samples in actual surveys because of the need for a large sampling frame. Hence, it is used only if the target population is small, such as drawing randomly from a party hat. For large populations, we need to find other ways to draw our sample.

Systematic sampling

In systematic sampling, we first divide the target population into smaller lists and then draw the elements following a particular pattern after a random start. For instance, if a town has a phone directory comprising 500 pages and the desired sample size is 1,000, then we can draw two elements from each page, such as the 10th and 50th names. The 10th and 50th positions are decidedly randomly. For example, we could have drawn the 20th and 40th names from every page. Hence, systematic sampling is similar to simple random sampling because we use random selection to draw the elements.

Stratified sampling

Unlike simple random sampling and systematic sampling, stratified sampling is more commonly used in actual surveys. Here, we first stratify the target population according to some criteria such as gender, income, workplace, residence, or cohort. We then draw the elements *randomly* from each stratum.

We may draw our samples from each stratum proportionally or disproportionately. In proportional stratified sampling, we draw the samples in equal proportion. For instance, if we have a class of 40 boys and 60 girls, we may draw 20% of the students from each category, that is, 8 boys and 12 girls.

In disproportionate stratified sampling, the sample proportions are unequal. For example, we may stratify contractors according to their contracting ability, such as, grades A, B, and C contractors. Because there are fewer of the large grade "A" contractors, it makes sense to select all contractors in this category. In contrast, because there are many small grade "C" contractors, it suffices to select a small sample from this stratum. Compared to the larger grade "A" contractors, these small contractors are more homogeneous. Hence, a large sample is not necessary. Often, there is little value in selecting similar sampling elements. We learn more by studying dissimilar cases.

It is also possible to stratify a sample using a combination of variables. In Table 5.1, the students have been stratified by class (Year 1 to Year 4),

Table 5.1 Sample of students stratified by class, religion, and gender.

	Population				Sample			
	Males		Females		Males		Females	
Religion	Yes	No	Yes	No	Yes	No	Yes	No
Year 1	30	30	30	30	3	3	3	3
Year 2	35	35	35	35	4	3	4	3
Year 3	40	40	40	40	4	4	4	4
Year 4	45	45	45	45	4	5	4	5
Total	150	150	150	150	15	15	15	15

whether they belong to a particular religion (Yes or No), and gender. We use proportional stratified sampling to select about 10% of the sampling units in each category.

Cluster sampling

In cluster sampling, we select random clusters rather than elements. For example, if we want to interview a sample of 200 students living on campus, we may select the students individually using simple random sampling. However, this is a tedious process because it requires a sampling frame consisting of the names of every student living on campus. Another disadvantage with simple random sampling is that we then need to interview the selected students in different places.

A more efficient strategy is to use multi-stage cluster sampling. For example, if there are 7 hostels on campus, we first select 4 hostels randomly. In the second stage, we select 50 students from each of these 4 hostels. Notice that the sampling frame has been drastically reduced from a long list of student names to a list of 7 hostels.

Cluster sampling is often used in city surveys to identify potential respondents. In the first stage, we select a sample of suburbs from a sampling frame of suburbs. In the second stage, we select different neighborhoods from these selected suburbs. In the final stage, we select potential respondents living in apartment blocks or along particular streets in each neighborhood. In each stage, we draw the elements randomly. This is why cluster sampling is a probability sample.

Non-probability samples

If a sampling frame is not available or too difficult to construct, then non-probability samples may be used. Chance selection is not used in non-probability samples because of the lack of a sampling frame. Hence, the probability of an element being selected is unknown. However, non-probability samples are easier to collect, which explains their popularity despite the higher probability of bias.

Non-probability samples include:

- convenience samples;
- purposive samples;
- quota samples; and
- snowball samples.

Convenience sampling

In convenience sampling, we select the elements out of our convenience or because we think they are likely to be good respondents. It is useful for exploratory work, for the pretesting of questionnaires, or where a quick opinion is required. Reporters often use this sampling technique for the evening news. It is also widely used in mall intercepts, street surveys, or email surveys.

In these cases, the researcher is more interested in some quick responses than in representativeness. Because the responses depend on whom you have selected, the sampling errors may be large.

Judgmental sampling

Judgmental sampling is purposive, that is, our judgment or the choice of experts is preferred to random sampling. An example of when we may prefer to use our judgment is in the construction of the Consumer Price Index (CPI). Here, we deliberately choose the sample basket of goods and services used to compute the CPI. The basket should represent a typical consumption pattern.

Obviously, the sampling errors depend on the quality of judgment. Different experts may not agree on what is representative.

Quota sampling

A quota sample is similar to a stratified sample except that *chance selection is not used* in each stratum. Hence, a quota sample is not a probability sample. In the hostel example, we may select the 50 students from each of the 4 hostels out of convenience rather than draw them randomly from a sampling frame.

Quota samples are popular because stratification reduces the need to select large samples. The greater homogeneity of the sampling elements with each stratum also reduces the sampling error. Finally, a sampling frame is not required, which makes it cheaper to draw a sample.

Snowball sampling

A snowball sample begins with a few respondents who provide referrals for the researcher to contact additional respondents. This may happen if the initial sample is very small, such as if we are sampling people who suffer from a rare disease or corporate leaders who are harder to access.

Sample Size

There is no simple guide on sample size. A misconception is that the sample size for the hostel survey (for example) should be 30 because it is where the Student *t* distribution is nearly a normal curve. This way of determining the sample size is incorrect. The right way of thinking about sample size is to link it to *statistical power* (Cohen, 1988), and this is discussed below.

In Fig. 5.2, the left curve is the distribution of the test statistic under the null hypothesis (H_0) and the right curve is its distribution under the alternative hypothesis (H_1). It is worth reiterating that a *statistical* hypothesis is about the values of a parameter, such as H_0: $\mu = 3$ and H_1: $\mu = 6$. It differs from a *research* hypothesis, which is about causal mechanisms. The two types of hypotheses are related because a causal mechanism may suggest certain values of a parameter.

In hypothesis testing, we can make two errors:

- Type I (α): Reject H_0 when it is actually true; or
- Type II (β): Accept H_0 when it is actually false.

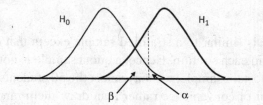

Fig. 5.2 Statistical power.

Table 5.2 Type I and Type II errors.

	H_0 is actually	
Decision	True	False
Reject H_0	Type I error	Correct
Accept H_0	Correct	Type II error

Table 5.3 Court analogy of Type I and Type II errors.

	Person is actually	
Decision	Innocent	Guilty
Guilty	Type I error	Correct
Innocent	Correct	Type II error

It may be clearer to view these errors in tabular form (Table 5.2). The court analogy is also helpful (Table 5.3). Here H_0 is the hypothesis that the person is innocent. The jury may send an innocent to jail (Type I error) or to set a criminal free (Type II error). As to which error is more serious is a matter of opinion. Generally, a Type I error is considered more serious, because most people do not wish to have a justice system that sends innocent people to jail or worse, face the death penalty.

From the perspective of quality control in manufacturing, H_0 is the hypothesis that the batch is good. Rejecting a good batch by mistake (Type I error) is probably less serious than not rejecting a bad batch (Type II error), which can seriously damage the reputation of the firm and result in loss of customers.

Usually we fix the Type I error α (for example, $\alpha = 0.025$) and obtain the critical value from statistical tables to draw the vertical dashed line in Figure 5.2. For simplicity, we assume the statistical distributions are normal. The value of α is also called the significance level. A value of $\alpha = 0.025$ for one tail or 0.05 for two tails of the normal distribution means that there is a 1 in 20 chance of making a Type I error. As a simple example, suppose John claims that he is able to tell, from a person's handwriting, whether that person is male or female. If we line up 20 handwritten letters for John to test his predictions, how many times must John guess correctly to persuade us that he has the ability to link handwriting to gender? If John makes a random guess, he will get, on average, 10 correct and 10 incorrect answers. Even if John gets 15 correct and 5 incorrect answers, we may think it is just fluke. By convention, scientists think John has the ability if he can guess 19 out of 20 cases correctly. In other words, he makes an error of 1 in 20, or 5%.

The power of a test is given by $1 - \beta$. A test of high power implies that β should be as small as possible. If β is large, the Type II error, which is β itself, will be large. As seen in Fig. 5.2, β will be reduced if:

- the spread of each curve is smaller (called spread size); and
- the curves are further apart (called effect size).

This implies that we should aim for greater precision (that is, smaller spread) and larger effect size in our samples. For example, in a survey on household consumption expenditure, we may stratify our population by household income to reduce the heterogeneity within each income group.

A larger effect size implies that the difference between the sample means should be large. In terms of Figure 5.2, the two curves should be further apart, that is, the two samples differ substantially with respect to the responses expected from the survey. By way of analogy, effect size is similar to the way we test a drug in an experiment. If the effect of the drug is weak, we will not be able to detect any difference in the mean responses between the experimental and control groups.

There are important caveats to the above rules. First, they only refer to a single variable. If there are many variables, the rules are not applicable. In most surveys, there are a large number of variables, so the rules are not directly applicable. Second, and more importantly, if we are seeking qualitative responses, we should select samples for varied, rather than similar, responses. To put it in another way, statistical precision may not be a useful criterion to qualitative researchers.

A final determinant of sample size is of a pragmatic nature. It has to do with access to potential respondents or the cost of doing so. If they are hard to reach, it will affect the sample size.

In summary, we use different criteria to determine sample size. In qualitative surveys, the aim should be to gather as many differing views as possible. In quantitative surveys, the spread and effect sizes come into play, together with the cost or difficulties in accessing potential respondents.

Errors in Surveys

Survey errors consist of non-sampling and sampling errors. *Non-sampling errors* include:

- *administrative errors* arising from mistakes in data collection or processing such as talking to the wrong person or keying in the wrong data;
- *respondent errors* in giving incorrect answers because they cannot recall, wish to hide information, are not in the mood to talk, or are put off by the questions;
- *conceptual errors* on the part of the researcher, such as the use of different and confusing concepts of culture; and
- *measurement errors* that occur because the measure of a concept is inappropriate, such as using the level of education as a measure of productivity.

Sampling errors arise from *chance* variations such as the variation of sample means from the population mean. Even if we draw the representative samples randomly, there will inevitably be sampling errors. For example, if we take representative samples of students from different classes in the same grade to compute the average weight, the sample means will vary slightly from class to class. From the Central Limit Theorem in statistics, the size of the variation is $\sigma/\sqrt{n}$, where σ is the population standard deviation and n is the number of students sampled in each class.

Pilot Survey

We often conduct a pilot survey involving a small but similar sample to determine if the survey design and method of data collection (for example, a questionnaire) may be improved before the actual larger-scale survey. Logistical and other field issues may also surface during the pilot study.

A pilot survey should always be conducted because there will always be areas for improvement before conducting the full survey. It may not be necessary for a well-tried and tested survey, unless there are new questions or procedures.

Perceptions of Train Service

Our example of a survey research is the design of a perception survey among commuters of the services provided by a mass rapid transit (MRT) system. It will be an annual trend study to track changes in perceptions of the quality of services over time. Train operators often carry out such surveys to obtain feedback from commuters to improve their service (Jones and Stopher, 2003; BART, 2014).

These perceptions depend on the characteristics of the rater, such as income, age, and gender. The quality of train services includes:

- the ease of access to train stations,
- the degree of integration with bus services,
- the ease of buying tickets,
- the fare structure,
- the extent of overcrowding,
- waiting time,
- cleanliness of station,
- station shopping experience,
- helpfulness of service staff,
- safety,
- security,
- internal carriage temperature,
- cleanliness of trains,
- quality of wireless connections,
- response to breakdowns, and
- the usefulness of commuter information.

Table 5.4 Quota sample of commuters per station.

Age group	Males	Females
20–39	10	10
40–59	10	10
60 and over	10	10
Total	30	30

A simple 5-point rating scale is used to rate each attribute of service.

There is no sampling frame because of the difficulties in generating a list of names for commuters. Hence, non-probability sampling is used. If there are 40 stations in the train network, we will select 60 commuters per station during the morning peak hours, resulting in a sample size of 2,400. We will use a quota sample stratified by gender and age group (Table 5.4). Income is not used for stratification because this variable is harder to observe during peak hours.

Alternatively, it is possible to pick 20 stations randomly and select 120 commuters per station for the study. The sample size is still 2,400, but this sampling strategy is more efficient to implement. However, not all stations are covered.

We will carry out a pilot study involving two stations to obtain feedback on the questionnaire and field organization.

References

BART (2014) *2014 Customer satisfaction study*. San Francisco: BART Marketing and Research Department.

Cohen, J. (1988) *Statistical power analysis for the behavioral sciences*. London: Routledge.

Jones, P. and Stopher, P. (2003, Eds.) *Transport survey quality and innovation*. London: Emerald.

Hsiao, C. (2003) *Analysis of panel data*. Cambridge: Cambridge University Press.

Wirth, L. (1938) Urbanism as a way of life. *American Journal of Sociology*, **44**, 1–24.

CHAPTER 6

Research Design III: Comparative Design

Features of Comparative Designs

A comparative study compares and contrasts several cases to draw inferences. This overcomes a weakness of causal case study design, which does not allow for the use of comparisons to draw causal inferences.

In terms of sample size, a comparative design is a small-N design. It falls somewhere between a case study and a survey. Hence, it is attractive to researchers who wish to explain phenomena where there are only a few cases. For this reason, comparative studies are widely used in political science to study social revolutions in a few selected countries (Paige, 1975; Skocpol, 1979). The small sample size makes it difficult to use multivariate statistical analyses or broad large-sample surveys.

In cultural studies, Parker (2001) compared the police system in Japan and America to account for the different crime rates. Brake (1985) compared youth culture in America, Britain, and Canada. Zippel (2006) studied differences in sexual harassment in the United States, European Union, and Germany.

Comparative studies are intended to be causal, that is, the researcher deliberately chooses the cases to discover causality, rather than through random sampling. For example, many researchers use the comparative approach to explain the economic performance of the East Asian "dragon" economies (Vogel, 1991; Schuman, 2010; Perkins, 2013). Klein (1988) compared slavery in the two New World colonies of Cuba and Virginia. Both colonies were plantation economies with similar proportions of slaves in the population. He used institutional differences to explain why slaves in Cuba achieved greater social integration and occupational mobility.

Despite the variety of applications, comparative designers use two main strategies in their quest to determine probable causes, namely, the Method of Agreement and the Method of Difference (Mill, 1884). The Method of Agreement is widely used. We discuss these designs below.

Method of Agreement

In the Method of Agreement, we look for common factors that are likely to cause the same outcome. For example, suppose Joe and Jim went to a lunch party and both had food poisoning (Table 6.1).

In the Method of Agreement, we start with the *same* outcome, food poisoning. Then we look for the *common* factors that may explain the event. In this case, both boys ate Dishes A and B, so we conclude that these two dishes are the likely causes.

Method of Difference

In the Method of Difference, we start with *different* outcomes and look for cases that are as similar as possible. If a factor is different, it is a probable cause.

Using the same food example, both boys went to a party and ate some food. Hence, their cases are similar: they were at the same party. Now suppose that Joe had food poisoning but Jim did not, that is, the outcomes are different (Table 6.2). We then look for a factor that is *different* between the two cases. In this example, Dish C is the probable cause.

Table 6.1   Example of method of agreement.

Dish	Joe	Jim
A	1	1
B	1	1
C	0	1
D	0	0
E	1	0
Outcome	Food poisoning	Food poisoning

Table 6.2 Example of method of difference.

Dish	Joe	Jim
A	1	1
B	1	1
C	1	0
D	0	0
E	1	1
Outcome	Food poisoning	No food poisoning

Comparison and Causation

Common factors do not prove cause. For example, one can drink water and a type of liquor each day of the week and get drunk. The only common factor here is water, and it cannot be the cause of drunkenness.

In the first food example above, there is no causal mechanism linking Dishes A and B to food poisoning. At best, they are probable causes. The food poisoning may have been caused by something else they had taken during dinner. Similarly, in the second food example, there is also no causal mechanism to link Dish C to food poisoning. Perhaps Joe had eaten something else after the party, and it caused the food poisoning.

In both cases, there is a correlation between eating dishes and food poisoning, but it does not prove cause.

Sampling

If correlation does not prove cause, we have to select our cases properly to make a more persuasive argument. A basic principle is to select diverse or dissimilar cases. For example, if we are trying to identify a small set of key variables that explain economic performance, then we should select countries that differ in these variables. For instance, if we consider natural resource endowment as a key variable (Barbier, 2007), then we should select countries with varied resource endowments. This may vary from resource-poor countries such as Singapore, to resource-rich countries such as Canada.

A second principle in comparative sampling is to select both successful and unsuccessful cases. A common mistake is to select only successful cases for comparison to identify a set of common factors. For example, some books on project management (e.g., Camilleri, 2011) tend to stress on the "critical success factors." The problem here is that the common factors may also be present in unsuccessful cases.

Finally, we should be aware of reverse causality in comparative studies. For example, while it is common to assume that countries that are rich in natural resources have a head start in economic development, it can also be a curse (Menaldo, 2016), the so-called "Dutch disease" where vast natural gas discoveries in 1959 led to a sharp rise in the Dutch guilder in the 1970s. This made other sectors of the Dutch economy less competitive, resulting in a rise in unemployment. The Dutch case is a clear refutation of any simplistic link between natural resource endowment and economic growth. It also illustrates the need to consider time adjustment, that is, whether we are looking at short run or long run events.

Four Little Dragons

Our selected example is Vogel's (1991) comparative study of the spread of industrialization in East Asia from the 1960s to the 1980s, in the so-called "four little dragons" of South Korea, Taiwan, Singapore, and Hong Kong. It is one of many studies of the rapid growth of East Asian economies (e.g., World Bank, 1993).

Vogel identified nine factors that may explain the rise of East Asia, namely:

- massive external aid from the US and international agencies;
- destruction of the old order and its replacement by a "strong State" that is relatively insulated from the local elites;
- sense of political and economic urgency for industrial development on the part of its leaders and the community;
- eager and plentiful labor force;
- familiarity with the good performance of the Japanese model of economic development, beginning with labor-intensive industries, export growth, skills and technological upgrading, and the crucial role of the Developmental State in guiding these changes;
- culture, such as a respected and meritocratic bureaucracy, an entrance examination system that rewards learning and discipline, group cohesion, and self-cultivation and learning;

- consumerism, the passionate drive to acquire new goods as incomes rise;
- export orientation; and
- the success cycle, where successful industrialization develops confidence to acquire new skills for another round of achievement.

To make comparisons easier, it is helpful to reduce the number of variables to a more manageable set. The massive external aid after World War II is a one-off event and we can remove this factor from consideration. We will also remove the urge to industrialize as it is present in most developing countries. Most developing countries prefer industrialization to agricultural development to expedite the creation of jobs and shift away from excessive reliance on commodity production.

For the cultural factors, we can combine the first two factors and rename it as having an effective bureaucracy. Nowadays, groupism is no longer viewed as particularly East Asian; on the contrary, the rational individual is more common, and groupism is merely *constructed*, such as in the literature on organization culture. Groupism also has other disadvantages, such as collective myopia (Chikudate, 2015). Finally, self-cultivation is also not a primarily East Asian trait. For example, philosophers, novelists, theologians, and poets wrote at length about self-cultivation in Germany (Bruford, 2010).

For the last three factors, consumerism appears to be an effect of rising income rather than a cause of growth. Similarly, the success cycle kicks in only *after* the economy has been set in motion. Hence, we are left with only export orientation as a possible cause of rapid growth in East Asia.

In summary, we have the following five common factors:

- destruction of the old order;
- its replacement by a Developmental State and effective bureaucracy that is relatively insulated from the local elites;
- a disciplined and plentiful labor force;
- export orientation; and
- familiarity with the Japanese model of development.

These factors have also been identified in the Development State literature (Johnson, 1982; Amsden, 1989). Here one may add other factors such as market-conforming interventions, active consultations with the private sector, industrial targeting, becoming an entrepreneur as a last resort, having a pilot agency, and the ability to resist rent seekers and discipline workers and poor corporate performers. The term "soft authoritarianism" is sometimes used to describe such Developmental States.

However, are the above set of factors the formula for economic success in East Asia? There are doubts. For neoclassical economists, the East Asian miracle is a myth (Krugman, 1994) because it is largely driven by "brute force" growth of capital and labor inputs rather than by total factor productivity growth. In other words, much of it was perspiration rather than inspiration.

In recent years, intense international competition, rising land and labor costs, as well as slower or falling growth of the labor force have reduced economic growth in these countries (Kozul-Wright and Rayment, 2007). Exports have also fallen or grown far more slowly, in line with the slower growth in global trade over the last decade (Hoekman, 2015). From this perspective, the high growth rates from 1960s to the 1980s in East Asia is a one-time event that is unlikely to be repeated. The future is more uncertain, and it is difficult for the State, or anyone else, to pick winners.

From a different perspective, the long post-World War II boom may have led to the development of powerful distributional coalitions (interest groups) that slow down a country's capacity to respond to structural changes, and thereby reduce the rate of economic growth (Bates, 1981; Olson, 1982).

Finally, many other States have imitated the industrialization strategies of the East Asian dragons, and yet failed miserably or achieved only modest economic growth (Mkandawire, 2001). In other words, the common factors, in different configurations, also exist in failed cases. Development is not just about State policies and capacities; performance also depends on external factors.

Limitations of Comparative Designs

We have seen that a comparative design may be used to identify a set of common factors. However, there are several shortcomings, such as:

- the inability to prove cause;
- difficulties in finding suitable, diverse cases;

- difficulties in finding comparable cases;
- difficulties in sorting out rival explanations if there are many variables and only a few cases (Lijphart, 1971); and
- the outcomes are binary.

It is possible to address some of the shortcomings. For example, in Paige's (1975) study of the effect of agricultural exports on the social movements of cultivators, he excluded the oil producing countries and small city-states without monocrop economies. This means that the theory is also less general.

Comparative studies with binary or dichotomous outcomes involve the presence or absence of causes or effects — it is all or nothing. In practice, it is usually a matter of degree, not of kind. For example, in our first food poisoning study example, a person may have eaten only a spoonful of dish A, and did not have food poisoning. There have been attempts to overcome this limitation by using truth tables and fuzzy or Boolean logic, called Qualitative Comparative Analyses (Ragin, 1987), with limited success.

References

Amsden, A. (1989) *Asia's next giant: South Korea and late industrialization.* Oxford: Oxford University Press.

Barbier, E. (2007) *Natural resources and economic development.* London: Cambridge University Press.

Bates, R. (1981) *Markets and States in tropical Africa.* Berkeley: University of California Press.

Brake, M. (1985) *Comparative youth culture: The sociology of youth cultures and youth subcultures in America, Britain, and Canada.* London: Routledge.

Bruford, W. (2010) *The German tradition of self-cultivation.* Cambridge: Cambridge University Press.

Camilleri, E. (2011) *Project success: critical factors and behaviors.* London: Routledge.

Chikudate, N. (2015) *Collective myopia in Japanese organizations.* London: Palgrave MacMillan.

Hoekman, B. (2015, Ed) *The global trade slowdown: A new normal?* London: CEPR Press.

Johnson, C. (1982) *MITI and the Japanese miracle: The growth of Industrial policy, 1925–1975*. Stanford: Stanford University Press.

Kozul-Wright, R. and raiment, P. (2007) *The resistible rise of market fundamentalism*. London: Zed books.

Krugman, P. (1994) The myth of Asia's miracle. *Foreign Affairs*, **73**(6), 62–78.

Lijphart, A. (1971) Comparative politics and comparative method. *American Political Science Review*, **65**, 682–693.

Menaldo, V. (2016) *The institutions curse: Natural resources, politics, and development*. London: Cambridge University Press.

Mkandawire, T. (2001) Thinking about developmental States in Africa. *Cambridge Journal of Economics*, **25**(3), 289–313.

Mill, J. (1884) *A system of logic*. London: Longman.

Olson, M. (1982) *The rise and decline of nations*. New Haven: Yale University Press.

Paige, J. (1975) *Agrarian revolution*. New York: Free Press.

Parker, C. (2001) *The Japanese police system today: A comparative study*. London: Routledge.

Perkins, D. (2013) *East Asian development*. Massachusetts: Harvard University Press.

Ragin, C. (1987) *The comparative method: Moving beyond qualitative and quantitative strategies*. Berkeley: University of California Press.

Schuman, M. (2010) *The miracle: The epic story of Asia's quest for wealth*. New York: Harper Business.

Skocpol, T. (1979) *States and social revolutions: A comparative analysis of France, Russia, and China*. Cambridge: Cambridge University Press.

Vogel, E. (1991) *The four little dragons*. Cambridge: Harvard University Press.

World Bank (1993) *The East Asian miracle*. Washington DC: World Bank.

Zippel, K. (2006) *The politics of sexual harassment: A comparative study of the United States, the European Union, and Germany*. London: Cambridge University Press.

CHAPTER 7
Research Design IV: Experiment

Features of Experimental Design

Suppose the relation between causes (Xs) and an effect or outcome (Y) is given by the hypothesis

$$Y = f(X_1, \ldots, X_k)$$

where Y is the dependent variable and $X_1, \ldots, X_k$ are k independent variables. The function $f(.)$ specifies how Y and Xs are related, but it may be unknown or imprecisely known. It is a causal mechanism linking the Xs to Y.

We use an experimental design if k is small and the possibility exists for *manipulating* some variables to ascertain their effects on the Y by keeping other Xs constant. For example, we may have $k = 5$ independent variables and we are interested to see how X_1 and X_2 affect Y, keeping X_3, X_4, and X_5 constant. We call the variables X_1 and X_2 *treatments* because of the frequent use of experiments in clinical and agricultural studies.

Many researchers think experiments provide stronger evidence of causation by eliminating rival explanations through experimental control. However, this is only possible if there are a few variables. If k is large, it becomes more difficult to disregard the influences of other independent variables by fixing their values.

Among social scientists, psychologists may conduct experiments in laboratories to observe how people respond to incentives (Abelson *et al.*, 2003). Other social scientists tend to undertake field experiments to evaluate policy (Orr, 1998; Dunning, 2012). However, social experiments are not common because k is large and it is often difficult to manipulate or fix socio-economic variables. For example, the economist cannot manipulate interest rates to observe its effects on business investment. All he can observe are two time series, one for interest rate, and the other for business investment. At most, he can regress or correlate one series with another. Hence, the economist conducts observational, rather than experimental, studies.

There are many types of experimental designs (Dean and Voss, 1999; Box *et al.*, 2005; Montgomery, 2009). We discuss the basic designs below.

Classical Experimental Design

In a classical experimental design, an experimenter uses two groups of subjects, called the *control* group (C) and the *experimental group* (E). He then administers a *treatment* (T) to E, and C provides a baseline to compare the effects:

$$
\begin{array}{cccc}
E & X_E & T & Y_E \\
C & X_C & & Y_C
\end{array}
$$

The treatment may be a drug, a new teaching technique, a new material, or a new scheme for financing cars. For simplicity, we assume that both groups are of equal size (n), although it is unnecessary as long as they are comparable (Table 7.1).

In a *blind experiment*, the subjects in each group do not know whether they have been given the treatment or placebo (inactive substance) because both pills look identical. The purpose of a blind experiment is to reduce the possibility that a subject may exaggerate the main or side effects if he knows that he has been given the active drug.

It is also possible to blind the investigators as well in a *double-blind experiment*, that is, they do not know whether the results come from E or C. Again, the purpose is to reduce the possibility that investigators are biased towards what they are looking for, such as whether the drug is likely to work.

If T is effective, then $Y_E - X_E$ will be greater than $Y_C - X_C$ within limits of experimental errors. In Table 7.1, X and Y are the pre-test and post-test scores, such as the mathematics test score of students in a class before and after introducing T. Subjects in the control group C did not receive the treatment, that is, they were taught using the existing method.

Table 7.1 Classical experimental design.

Student	Pretest	Treatment	Post-test
1	90		95
2	80		70
3	60	*T*	80
...			
n	70		75
1	60		65
2	70		75
3	85	*Nil*	80
...			
n	90		85

In some cases, it may be difficult to conduct the experiment and some adjustments have to be are made, such as:

- by not using a control group, perhaps because it is difficult to find another group large enough for the experiment;
- by not conducting a pre-test, perhaps because it is difficult or expensive to conduct, or because the groups are large enough for the experimenter to assume that they are comparable; or
- by discarding the pre-test scores in the data analysis, possibly because the pre-test scores are used primarily as a check that both groups are comparable before the treatment is administered.

Campbell and Stanley (1963) called these departures from the classical experimental design *quasi-experimental designs*. Obviously, conclusions from quasi-experimental designs are less persuasive than the full classical experimental design.

Nonetheless, there are several *threats* to the classical experimental design, such as:

- subjects maturing during the experiment, which may be handled by keeping the period between pre-test and post-test short;

- measurement error, where the mathematics tests are inappropriate, such as if they are too difficult and cheating occurs;
- the testing effect, where subjects become better because they learn to do better after completing the pre-test, or they have been "taught to the test," that is, they have been drilled to do well;
- subjects reacting differently because they know they are being observed by taking part in an experiment (Hawthorne effect);
- an external event, such as a change of teacher teaching the control group, which may invalidate the results;
- if sample sizes are too small to provide firm conclusions, which can be increased by adding more subjects in *E* and *C* (i.e. through *replication*) or by using more than two groups; and
- if the groups are not comparable.

Researchers use a variety of techniques to ensure that the experimental and control groups are comparable. These strategies include:

- randomization;
- matching; and
- repeat measures design.

The *random allocation* of subjects to *E* or *C* will tend to cancel out systematic differences between the groups if the sample sizes are sufficiently large. A simple example of a random allocation is to assign a number to each subject and then draw the numbers from a hat to ensure two roughly equal groups. Many other simple allocation methods exist, such as flipping a coin or drawing from a bag with two colors.

The experimenter may also *match* the students on mathematical ability by generating a ranked list of student names according to their previous mathematics grade. Students with odd ranks are than assigned to *E* and those with even ranks are assigned to *C*. This is similar to the way we assign people to either side of a tug of war game.

In a *repeat measures design*, we use only one group of subjects for the experiment. However, we measure each student twice, once using the traditional method and another time using the new teaching technique. In other words, each student acts as his own control, which means that the variability in scores *between* students do not matter. We will explain why this is the case shortly, in our example for Table 7.5.

Among these three methods, random allocation is probably the easiest to carry out. However, there are instances where the sample size is small, such as a village school with small class sizes. Random allocation or matching may then not result in comparable experimental and control groups. The experimenters may try using a repeat measures design.

However, random allocation, matching, and repeat measures designs do not solve the problem of sampling bias through *self-selection*, that is, the participants in the experiment may not be representative of the wider community. This may happen for various reasons. For example, if monetary incentives are given, it may attract poorer subjects who have higher marginal utility for money.

Parallel Group Design

A parallel group design is similar to the classical experimental design except that the second group now receives a treatment, say U (Table 7.2). In our example on the teaching of mathematics, the treatment may consist of a second way of teaching mathematics. In the case of clinical trials, the second treatment may be a different drug or a different dosage of the same drug. This design faces the same threats as the classical experimental design.

Table 7.2   Illustration of parallel group design.

Student	Pretest	Treatment	Post-test
1	90		95
2	80		70
3	60	T	80
...			
n	70		75
1	60		70
2	70		70
3	85	U	80
...			
n	90		95

Repeat Measures Design

Recall that a repeat measures (or paired) design may be used to deal with the problem of having two groups that are not comparable. Here, only one group with sample size N is used and each student is taught in two different ways (T and U), one after another with a sufficient gap between the two periods (Table 7.3).

As a second example, consider a street experiment to determine if wine A or B tastes better. In a parallel group design, two groups of people are used, one for each wine, and we obtain the ratings for each wine on a scale of 1 to 10 (Table 7.4). There is no pre-test rating. Note the large

Table 7.3 Illustration of repeat measures design.

Student	T	U
1	95	90
2	80	75
3	60	85
...		
N	75	70

Table 7.4 Parallel group design in wine tasting experiment.

Participant	Wine	Rating
1		9
2		7
3	A	4
...		
n		7
1		3
2		7
3	B	8
...		
n		5

Table 7.5 Repeat measures design in wine tasting experiment.

Participant	Wine A	Wine B	d
1	8	9	1
2	4	5	1
3	6	8	2
...			
N	7	7	0

variability in ratings among the participants. It is difficult to obtain two groups that are similar in tastes.

In Table 7.5, we show a repeat measures design for the wine tasting experiment. Notice that the sample size is now smaller, which means this design is cheaper to implement. Also, observe that some wine tasters are higher scorers (for example, Participant 1), while others are low scorers (for example, Participant 2). Yet, this variability *between* participants does not matter, because we are only interested in the difference in ratings (d) *within* each participant. This is the main advantage of a repeat measures design. There is less variability in the data, which makes it easier to detect the difference in ratings.

Randomized Block Design

There is considerable evidence from past studies to show that boys tend to do better than girls in mathematics (Niederle and Vesterlund, 2010). This means that gender matters, and we now have two factors, a new way of teaching mathematics (T) and gender (G) that may affect mathematics scores. Here G is an extraneous variable because it is not of interest to the experimenter but its presence confounds the analysis. It is also called a "noise" variable, and the experimenter has to deal with the unwanted noise.

Suppose there are 100 students, comprising 60 boys and 40 girls, in a class. Previously, we assigned 50 students each to E or C through random allocation or matching without considering gender. If we wish to block out the effects of gender, we will need to randomly divide the boys into

Table 7.6   Illustration of blocking.

	60	30	E
100		30	C
	40	20	E
		20	C

two groups of 30 each, and similarly, we will divide the girls into two groups of 20 each (Table 7.6).

Observe that subjects within each *E* or *C group* are homogeneous, that is, they comprise all boys or all girls. In other words, the gender effect has been blocked out in this randomized block design.

References

Abelson, R., Frey, K. and Gregg, A. (2003) *Experiments with people: Revelations from social psychology*. New York: Psychology Press.

Box, G., Hunter, W. and Hunter, J. (2005) *Statistics for experimenters*. New York: Wiley.

Campbell, D. and Stanley, J. (1963) *Experimental and quasi-experimental designs for researchers*. Belmont, California: Wadsworth.

Dean, A. and Voss, D. (1999) *Design and analysis of experiments*. New York: Springer.

Dunning, T. (2012) *Natural experiments in the social sciences*. London: Cambridge University Press.

Montgomery, D. (2009) *Design and analysis of experiments*. New York: Wiley.

Niederle, M. and Vesterlund, L. (2010) Explaining the gender gap in mathematics test scores: The role of competition. *Journal of Economic Perspectives*, **24**(2), 129–144.

Orr, L. (1998) *Social experiments*. London: SAGE.

CHAPTER 8
Research Design V: Regression

Features of Regression Design

A regression design examines the *influence* of *independent* variables $(X_2, ..., X_k)$ on a *dependent* variable (Y). Symbolically, we write

$$Y = f(X_2, ..., X_k).$$

Here, $f(\cdot)$ denotes a function. Note that the subscript for X starts with 2 rather than 1 because of the presence of an intercept term, as shown in Equation (8.1). For notational brevity, we write it as $Y = f(X)$ if there is a single independent variable, and $Y = f(\mathbf{x})$ if there are many variables.

As an example, we may postulate that a household's consumption expenditure (Y) depends on its disposable income (X_2), and perhaps other factors such as household size (X_3), wealth (X_4), and interest rates (X_5). Similarly, in agriculture, we postulate that plant yield (Y) depends on the plant variety (X_2), the amount of fertilizer applied (X_3), the amount of water used (X_4), and other factors.

There is an important difference in the two examples. In the consumption case, the economist does not vary household income simply because he has to accept it as a given economic observation. He does not have the means to manipulate household income before observing its effect on consumption expenditure. In other words, he has observational rather than experimental data. In the agriculture case, the experimenter can apply different amounts of fertilizer and water to different plants and then observe the corresponding changes in plant yield. That is, he is able to manipulate changes in the independent variables and then observe their effects on Y. Another way of putting it is to say that he fixes the values of the different independent variables and then observes the corresponding changes in Y. We say that $\mathbf{x}$ is fixed in repeated sampling.

71

The relation between Y and X is asymmetric, that is, the dependent variable Y depends on changes in the independent variable X. This model does not explain changes in the independent variable X; this variable is *exogenous*, or is determined outside the model. The dependent variable Y is *endogenous*, that is, determined by the model or, specifically, by $f(\mathbf{x})$. Hence, regression differs from correlation, where the relation is symmetric without any specification on whether Y depends on X or vice versa. For example, we often say that a person's height correlates with his weight.

If we assume that $f(\cdot)$ is linear, we can write

$$Y = \beta_1 + \beta_2 X_2 + \cdots + \beta_k X_k + \varepsilon. \tag{8.1}$$

Here β_1 is the constant or intercept term, the other βs are parameters or slope coefficients, and ε is the error term. We assume that the errors are distributed normally with zero mean and constant variance, that is,

$$E(\varepsilon) = 0, \quad \text{and} \quad \text{Var}(\varepsilon) = \sigma^2.$$

Here $E[\cdot]$ denotes the expectation or mean, and $\text{Var}(\cdot)$ denotes the variance. Note that there are k parameters and $k - 1$ independent variables in the model.

If we have data on n households, we can write the model as

$$Y_i = \beta_1 + \beta_2 X_{2i} + \cdots + \beta_k X_{ki} + \varepsilon_i, \quad i = 1, \ldots, n. \tag{8.2}$$

For example, if we have data for consumption expenditure, income, and household size, then for the first household, the equation may be

$$1200 = \beta_1 + \beta_2 2000 + \beta_3 4 + \varepsilon_1.$$

That is, this particular household spends \$1,200 of its monthly income of \$2,000 on consumption, and it has a household size of 4. There are four unknowns, namely, the three βs and ε_1. For n households, we have n equations, one for each household. Usually, we will use Equation (8.1) rather than Equation (8.2) to avoid the tedious carrying of subscripts.

To estimate the parameters from these n equations, we usually use the method of least squares or *ordinary least squares* (OLS) to find the line of best fit

$$Y = b_1 + b_2 X_2 + \cdots + b_k X_k + e. \tag{8.3}$$

Do not mix up the *population* regression model in Equation (8.1) with the *sample* estimate in Equation (8.3). The bs are sample estimates of the βs, and e_i is a sample estimate of the error term ε_i. You may recall from elementary statistics that the sample standard deviation s is an estimate of the population standard deviation σ.

In OLS, we minimize the sum of squares of the residuals (e_i), and hence the terms "least squares" and "best" fit. That is, we minimize

$$\Sigma\, e_i^2 = \Sigma\, (Y_i - Y_i^*)^2.$$

The residual is the difference between an *observed* point (Y_i) and the corresponding *estimated* point on the regression line (Y_i^*).

The "ordinary" part of OLS refers to the standard classical least squares model, to differentiate it from more advanced models such as weighted least squares, generalized least squares, and nonlinear least squares. This estimating technique imposes several other important assumptions on the model to obtain the line of best fit, and we will discuss these issues in Chapter 14.

If we take the expectation in Equation (8.1), we have

$$E[Y] = \beta_1 + \beta_2 X_2 + \cdots + \beta_k X_k$$

because $E[\varepsilon] = 0$. Taking the partial derivative of $E[Y]$ with respect to any of the independent variables, say X_2, we see that

$$\partial E[Y]/\partial X_2 = \beta_2.$$

It may be clearer if we write it as

$$\Delta E[Y] = \beta_2 \Delta X_2.$$

That is, β_2 represents the effect of a unit change in X_2 on $E[Y]$, holding other variables constant. Hence, we can interpret regression as a form of *statistical control* where we hold other variables constant and consider the impact of each independent variable on Y. Unlike an experiment where we physically fix the values of extraneous variables to examine causality among a *few* variables of interest, it is not necessary in regression. By means of statistical control, we can determine the effects of *many* independent variables on the dependent variable. Thus, it is not surprising that regression analysis is popular in the social sciences where experimental control is difficult, if not impossible.

A regression model should specify the causal links between the independent variables and the dependent variable. It is not just a correlation exercise. There must be good reasons why certain variables are included in the model.

Sampling

In regression design, there are several considerations in sampling. The first consideration is the number of data points (n) relative to the number of parameters. In Equation (8.1), there are k parameters, and so the number of data points should be much higher than k. For instance, if there are two parameters, the regression equation becomes

$$Y = \alpha + \beta X + \varepsilon.$$

This is the equation of a straight line if we ignore the error term. Recall that the error term ε_i captures the deviation of an observed value Y_i from its corresponding point on the line, that is, for any ith data point,

$$Y_i - \varepsilon_i = \alpha + \beta X_i.$$

For any given *sample*, we estimate the regression line (see Figure 8.1). We can write

$$Y = a + bX + e \quad \text{or} \quad Y_i = a + bX_i + e_i$$

where a is the estimated intercept, b is the estimated slope coefficient, and e is the residual.

How many data points do we require to estimate the regression line of best fit properly? A rough rule of thumb is at least 10 points for each slope coefficient. In this example, there is only one slope coefficient (β) and we require a minimum of 10 data points. This rule is just a rough guide; in some cases, we simply do not have enough data points. If there are insufficient data points, we risk estimating an incorrect regression line.

The second consideration in sampling is to collect data that vary *substantially* in the X-direction, as shown in Fig. 8.1. This ensures that the estimated regression line will not be sensitive to any data point. For example, if we use only the cluster of five points near the middle

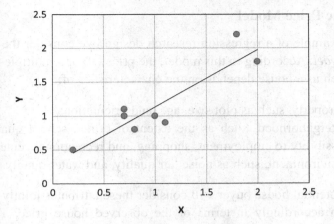

Fig. 8.1   Linear regression line.

of the graph, the estimated regression line will be nearly horizontal, which incorrectly reflects the upward-sloping linear relation between Y and X.

The third consideration in sampling is to collect representative samples. For example, suppose

$$Y = \alpha + \beta X + \varepsilon,$$

where Y is real Gross Domestic Product (GDP) or income per head, and X is a variable capturing a country's openness to trade. For example, X may be a country's average tariff rate or its trade dependency ratio, that is, the ratio of trade to GDP. We can collect cross-country data from, say, 20 countries and regress Y on X using these 20 data points. The *research* hypothesis is that countries that are more open to trade tend to enjoy higher incomes per head, because they will be able to benefit from trade in ways such as access to markets and technology, greater international competition, economies of scale, and learning economies. The related *statistical* hypothesis is that $\beta > 0$. The sampling issue here is that we may not have selected a representative sample of 20 countries. This issue arises because international trade may also impose adjustment and other costs (for example, declining terms of trade) so that the net effect is less certain.

Hedonic Price Model

As an example of a regression research design, we consider the *hedonic price model*. According to this model, the price (P) of a multiple-attribute asset such as a house depends on the characteristics of:

- the property, such as plot size, age, and orientation;
- the neighborhood, such as the extent of crime, school quality, and accessibility to employment, shopping, and recreational centers; and
- the environment, such as noise, air quality, and water quality.

A potential house buyer will consider these attributes jointly and pay for them accordingly in terms of the observed house price. Thus, for example, an accessible house will be more expensive than one in a remote location, holding all other characteristics equal. Hence, we can write the *house price function* as

$$P = \beta_1 + \beta_2 X_2 + \cdots + \beta_k X_k + \varepsilon. \tag{8.4}$$

The Xs are the property, neighborhood, and environmental characteristics of the house. Except for the intercept term, the βs are *implicit prices* (Rosen, 1974). These prices are not directly observable; for each property, we can only observe its transacted house price (P) and property characteristic. Our problem, then, is to estimate the implicit prices from a large sample of transacted prices.

For ease of estimation, we usually use a linear house price function. Typically, there are about 15 independent variables or property characteristics, so the rough minimum number of houses in the regression sample is 150 (Table 8.1). The variables are self-explanatory except for the noise variable, which is based on a 1 to 5 rating scale, with a rating of 1 as quiet and 5 as very noisy.

There are three main uses of an estimated hedonic price model. First, the property tax department may use it for mass appraisal for property tax purposes (Benjamin *et al.*, 2004). Here, we use the estimated implicit prices to estimate the value of *other* houses in the neighborhood that are not in the sample. Beyond the neighborhood, the technique may not be reliable because housing markets are local. House buyers in different cities may value each property characteristic differently.

Table 8.1 Data for hedonic price model.

House	Transacted price (P, in \$)	Land area (A, in m²)	Age (G, in years)	...	Noise (N, rating scale)
1	800,000	150	5	...	2
2	900,000	160	6	...	1
3	1,100,000	180	8	...	3
...	...	...	...	...	...
150	1,600,000	210	2	...	3

Second, developers can use implicit prices to build property features that are highly valued by house buyers and maximize the profit. Generally, buyers of new homes value tenure (for example, freehold or leasehold), land area, house design and conditions, security, amenities, and proximity to good primary schools (Ridker and Henning, 1967).

Third, we may use the estimated model as a surrogate (proxy) to value environmental quality such as noise or air pollution (Colony, 1967; Nelson, 1980; Nourse, 1967). There is no direct market for aircraft or traffic noise. However, the property market can serve as a surrogate (shadow) market. The implicit price is the shadow price. Shadow pricing is widely used in cost-benefit analyses for project appraisals.

The hedonic price model has other weaknesses apart from its limited ability to generalize beyond the neighborhood. First, the local housing market must be stable. Many housing markets go through booms and busts (King, 2010). If prices fluctuate too much, such as during a boom or bust, then the estimated regression equation will contain large errors. If house prices move erratically in the short run, it is difficult to estimate implicit prices with precision.

Second, buyers should have full information about the environmental characteristics of a property, such as soil or water contamination. They must also be able to evaluate their possible impacts on house prices so that they can adjust their bid prices accordingly (Bartke, 2011). If they do not have such information or are unable to evaluate them properly, the implicit prices for such environmental characteristics are unreliable.

Third, the linear equation should be a good approximation of the true relations between house price and its characteristics. If this is not so, such as if the relations are nonlinear, the estimated regression model will contain large errors. Hence, the choice of the function form for hedonic regression matters (Halvorsen and Pollackowski, 1981).

Fourth, we need to assume that data on housing characteristics, such as the land area, tenure, house condition, age, and amenities of each house, are available. Otherwise, it will not be possible to estimate the implicit prices without data. The large amount of data required is a major limitation of the hedonic price model.

Fifth, we need to specify the regression model properly so that we do not inadvertently omit important variables from the model. For example, if we exclude land area from the model, the estimated regression line may fit the data poorly because of the omitted variable bias (Abbott and Klaiber, 2011).

Sixth, we need to measure the variables correctly. For instance, is a simple rating scale appropriate as a measure for noise level? How do we measure the quality of housing design or its condition?

Seventh, potential buyers should be able to find houses with a bundle of characteristics that suit them. This may not be true. For example, it may not be possible to find a large, new house with a garden in the inner city area. Similarly, in the rental housing market, some potential renters may not be able to find suitable housing attributes.

Lastly, the local housing market must not be thin. If we do not have sufficient transactions during the sampling period, the observed house prices may not be reliable. This is less of a problem for houses. However, it can be a real problem for other types of properties. For example, there are fewer transactions for office buildings, factories, or shopping malls. Some researchers try to overcome this problem by using valuation prices as proxies for unobserved market prices.

The main lesson from this hedonic price model example is that regression analysis is not just a simple exercise of using statistical software to estimate the regression equation. There are many theoretical and practical considerations.

References

Abbott, J. and Klaiber, H. (2011) An embarrassment of riches: Confronting omitted variable bias and multiscale capitalization in hedonic price models. *Review of Economics and Statistics*, **93**(4), 1131–1142.

Bartke, S. (2011) Valuation of market uncertainties for contaminated land. *International Journal of Strategic Property Management*, **15**(4), 356–378.

Benjamin, J., Guttery, R. and Sirmans, C. (2004) Mass appraisal: An introduction to multiple regression analysis for real estate valuation. *Journal of Real Estate Practice and Education*, **7**(1), 65–78.

Colony, D. (1967) *Expressway traffic noise and residential properties*. Toledo, Ohio: University of Toledo Press.

Halvorsen, R. and Pollackowski, H. (1981) Choice of functional form for hedonic price equations. *Urban Economics*, **10**, 37–49.

King, P. (2010) *Housing boom and bust*. London: Routledge.

Nelson, J. (1980) Airports and property values. *Journal of Transport Economics and Policy*, **14**, 37–52.

Nourse, H. (1967) The effect of air pollution on house values. *Land Economics*, **43**, 181–189.

Ridker, G. and Henning, J. (1967) The determinants of residential property values with special reference to air pollution. *Review of Economics and Statistics*, **49**(2), 246–257.

Rosen, S. (1974) Hedonic prices and implicit markets: Product differentiation in pure competition. *Journal of Political Economy*, **82**, 34–55.

CHAPTER 9
Methods of Data Collection

Data Collection Methods

After determining the research *design*, the next step in the research process is to select the *methods* of collecting data. These methods include:

- observations;
- interviews;
- questionnaires;
- standardized tests;
- use of physical instruments;
- simulation; and
- review of documents.

Observations and interviews are widely used in qualitative research to collect contextual details and probe more deeply into a respondent's perspectives, interpretations, and experiences. Quantitative researchers tend to use questionnaires, standardized tests, physical measuring instruments, and simulation. Both types of research review existing documents for qualitative or quantitative data.

Many studies use multiple or *mixed methods* to collect data to exploit the strengths and offset the weaknesses of each data collection method (Blake, 1989). In doing so, they expand the scope of the research. For example, we may use standardized tests to assess the performance of students, followed by qualitative data on why students with similar backgrounds differ substantially in their performance. Economists regularly use such a strategy, where they supplement quantitative data with qualitative illustrative examples. For instance, Knoop (2015) provided quantitative data to test various theories of business cycles, followed by case studies of business cycles in a few selected countries or regions.

He considered the Great Depression and postwar business cycles in the US, the East Asian financial crisis of 1997–1998, the Great Recession in Japan (1992–2003), the Eurozone debt crisis (2008 onwards), and the global financial crisis of 2008–2009 that was triggered by the collapse of the US subprime mortgage market.

Instead of combining the qualitative and quantitative data, some researchers transform the qualitative data into quantitative data (Tashakkori and Teddlie, 1998). For example, a researcher may use numeric codes corresponding to the responses to open-ended questions in a structured questionnaire. This strategy partially overcomes the possibility of unlimited responses to open-ended questions.

Before discussing these data collection methods, we need to understand the scales from which empirical *measures* for theoretical concepts are developed.

Scales

Scales are used for categorization, ranking, and assessing magnitudes (Table 9.1). There are two types of variables, namely:

* *discrete* variables that take integer values, such as 20 girls; and
* *continuous* variables that take any real number, such as 2.3 kg.

A *nominal scale* categorizes data, such as 0 for females and 1 for males. In transport studies, the mode of transport to work is usually a categorical variable such as 1 for walk, 2 for bicycle, 3 for motorcycle, 4 for bus, and so on. We often use discrete variables to generate count or frequency data, such as the number of boys and girls in a class.

Table 9.1 Types of variables and scales.

Type of variable	Scales	Use	Example
Discrete	Nominal	Classification	Gender
	Ordinal	Ranking	Rating
Continuous	Interval	Distance	2005–2010
	Ratio	Ratio	Weight

An *ordinal scale* provides ratings, for example, 4 = Very Important; 3 = Important; 2 = Not So Important and 1 = Not Important. Another common technique is to provide a statement and respondents tick their answers depending on whether they strongly agree, agree, remain neutral, disagree, or strongly disagree with it. There is no hard and fast rule whether a 5-point, 7-point or 10-point scale should be used so long as one pays attention to the sensitivity of the responses. The difference between using odd and even numbers lies in whether "Neutral" is an acceptable answer. Technically, ordinal scales cannot be averaged because the intervals are unequal. However, if it can be reasoned that the intervals are roughly equal, then averaging may be acceptable.

An *interval scale* consists of equal intervals that measure the relative *distances* (differences) between points on the scale, such as IQ scores, temperature, or time. Ratios are meaningless, that is, a person with an IQ score of 160 is not twice as intelligent as one with a score of 80. Similarly, the period 2005–2010 makes sense in calendar time, but not the ratio 2005/2010.

In a *ratio scale*, the ratios are meaningful. For example, a length of 3 m is twice as long as 1.5 m, or 10 kg is twice as heavy as 5 kg.

Despite the differences in the interval and ratio scales, it is often not necessary to distinguish them. Both scales use real numbers, and most statistics such as the mean and standard deviation apply to them.

In general, how precisely we want to measure something depends on the purpose and cost. For instance, we should measure a room's temperature using a thermometer if this level of accuracy is important, rather than merely stating if it is "hot" or "cold." The scale of measurement will also affect the type of statistical tests that may be used. For instance, in the nominal scale, the mean and standard deviation are meaningless. If there are 20 boys and 10 girls in class, the average (15) does not make sense.

Observations

We may collect data by observing:

- the space or physical arrangements, such as the hierarchical arrangements of office space;
- the people, in terms of the number, composition, and their behavioral patterns;

- the goals, or what they are trying to accomplish;
- their actions;
- the activities;
- traces of frequency of use of equipment or facilities;
- flows of people or traffic;
- processes;
- occurrences, such as an accident;
- sequence of events; and
- feelings, such as whether people are happy, sad, or shocked.

Such observations may be independent or participatory. In the latter, the observer participates in the activities of the group. Such participation may be passive or active. In contrived observations, the observer may purposely lodge a "complaint" to see the reaction. The participant observer should try to "soak and poke" beyond appearances (Fenno, 1978).

The researcher must know what to observe. Most observers use a *checklist* to guide their observations. They also try to take notes and make sketches as soon as practical and supplement them with recording media. However, be sure to obtain permission to record or film conversations and activities. If immediate note-taking is not possible, it should be done at the earliest opportunity.

Observer bias may occur for various reasons. For example, the observer may either not understand the context or he did not receive adequate training. Any two observers may also interpret the same event differently. To reduce observer bias as well as to improve coverage, we often use two or more observers to compare notes. In general, *triangulation* refers to the use of two or more observers or methods to collect data.

Interviews

Most qualitative researchers use interviews in their research because of the opportunity to ask probing questions through conversations and open discussions. The participants may be representative of the population, key informants such as leaders or experts, or they form a group. The interviews may be unstructured, semi-structured, or structured. Some interviews require respondents to think aloud, such as to determine how students solve problems in educational research (Leighton, 2017).

In an unstructured interview, the interviewer does not wish to impose any prior framework. For instance, in an interview with a project manager on how he was affected by project failure, the interviewer may start with a question such as "how do you define project failure?" The interview then proceeds based on the responses. Generally, the questions will cover awareness of the issue, the adaptive responses, damage control, the consequences, and so on.

A semi-structured interview based on a research framework is more common. The interview has a list of questions, but the answers can be open-ended. Like structured interviews, researchers can use open-ended questions to probe more deeply into an issue.

A structured interview is highly standardized. Respondents answer the same set of questions in the same order. These structured questions contain a limited set of possible answers to ensure that aggregation and comparisons are possible between groups or across different periods.

A *focus group* is a group interview comprising about five to ten respondents. The researcher facilitates and moderates the collective discussion to explore ideas, share views, or make recommendations on an issue. Its effectiveness will depend on the composition of the participants, whether they are representative of the population, the skills of the moderator, the ground rules, and the type of questions.

If the group consists of experts, it is an *expert panel*. The *Delphi method* uses an expert panel to generate business forecasts over several rounds. The experts use the results from each earlier round to revise their forecasts. This assumes that group forecasts are more reliable than individual ones.

Beyond ten respondents, the focus group becomes a *community meeting*. Researchers use public forums and hearings to gather ideas from a wider range of stakeholders. Such meetings should be inclusive, that is, they include all sections of the community. They should also be participatory and not be dominated by certain stakeholders. Do not assume that all stakeholders will be present for a public meeting.

Researchers may carry out the interviews face to face, particularly if probing questions or visual aids are required, or over the telephone or Internet. Because the presence of the researcher may affect the respondent's readiness to provide information, the interviewer should try to put the respondent at ease.

Other considerations in interviews include authorizations, informant access to key respondents, timings, venues, seating arrangements, use of recorders (if permission is given), note-taking, and training of interviewers.

Questionnaires

A questionnaire is a list of questions intended to collect information by asking people directly. It is often used in conjunction with an interview. However, a postal, email or web-based questionnaire does not involve an interview.

Most questionnaires contain highly structured questions together with a limited set of answers. They usually contain factual questions and ratings, and occasionally opinions and reasons. An example is given below.

ABC Organization

Address

Contact details

Date

Dear Sir/Madam,

Study of tenant satisfaction

I am conducting a survey on tenant satisfaction to serve you better. Your response will be beneficial in helping us understand the areas we have done well, and the areas that need improvement. It would be appreciated if you could convey your views by answering the questions below.

Thank you for sparing your valuable time.

Sincerely,

Name and position

Note: If a covering letter is not used, a brief introduction outlining the purpose of the survey should be given in the questionnaire.

Please fill in the blanks or tick the appropriate box:

A Tenant information

1 Mall: _____   Unit number: _____

2 Name (optional): _____   E-mail address:_____

Contact number: _____   Job title: _____

B Tenant satisfaction

Please circle your *level of satisfaction* (1 = Low; 7 = High) for each listed item.

1 Lease management

    a. Lease period   1 2 3 4 5 6 7

    b. Rent collection   1 2 3 4 5 6 7

    c. Etc.

2 Marketing

    a. Marketing expenditure   1 2 3 4 5 6 7

    b. Marketing activities   1 2 3 4 5 6 7

    c. Etc.

3 Layout of shopping center

    a. Size of shops   1 2 3 4 5 6 7

    b. Shopper circulation   1 2 3 4 5 6 7

    c. Etc.

4 Etc.

Do you have any suggestions or comments?

....

....

....

End of Questionnaire. Thank you once again

Before finalizing the questionnaire, you should conduct a *pre-test* using a small sample of respondents to obtain feedback on the length, structure, sequencing, and content. Generally, a questionnaire should not exceed five pages. It is advisable to adopt a simple structure with proper sequencing without too many disruptive jumps from one section to another. Finally, check the content for *validity*. There are three types of validity, namely:

- construct validity, that is, are we measuring the right thing? For example, is it correct to measure satisfaction with lease period, rent collection, and so on as part of lease management?;
- internal validity, that is, is the causal mechanism that links causes to effects correct? For example, is it reasonable to postulate that lease management, marketing, layout of shopping center, and so on affect tenant satisfaction?; and
- external validity, that is, can the results be generalized to other contexts? For example, can we generalize the results of a tenant satisfaction survey from a local mall to malls in general, including malls in other countries?

After the pre-test, the questionnaire is then refined, such as by checking validity, proper numbering, rewording vague or offensive questions, including sufficient options in a question, removing duplicates and unimportant questions, and putting the more difficult or contentious questions last rather than first.

Some suggestions for improving a questionnaire include (Bradburn *et al.*, 2004; Brace, 2013; Harris, 2014):

- using simple words. If a technical word is necessary, provide a short explanation within the question; similarly, provide a map if necessary;
- using fixed-alternative questions that are theoretically sound and not artificially imposed, and state clearly if multiple answers to a question are possible;
- avoiding vague questions, such as what does "seldom" mean in terms of how many times I watch movies? It is better to provide frequency counts, such as "On average, how many movies do you watch in a month?";

- using open-ended questions where many answers are possible, for example, "What is your vision for downtown?";
- avoiding questions that lead to particular answers, for example, "Should *unproductive* speculators be taxed?";
- avoiding double-barreled questions, for example, "Is your work easy *and* challenging?" poses a dilemma if it is easy but not challenging;
- stating clearly the units of measurement, for example, gross or net monthly income;
- asking in units that people remember, for example, monthly take-home pay rather than annual income;
- using ranges for sensitive issues, for example, income ranges;
- de-sensitizing phrases, for example, "Many people surf the Internet for pornographic sites. Have you done this before?" is less sensitive than just "Have you surfed the Internet for pornographic sites?";
- avoiding hypothetical questions that tend to be poor predictors because there are many considerations, for example, "Do you intend to buy this product?";
- avoiding questions on competency, for example, "How do you rate yourself as a computer user?" is prone to the *prestige bias* of over-rating one's competency;
- avoiding questions that have a *social desirability bias*, for example, "Do you support this project to help the unfortunate children?"; and
- being aware of possible researcher bias, because the way questions are worded or *asked* may not reflect the way issues are *viewed* by respondents.

Sometimes, we carry out an *item analysis* during pre-test to determine if the responses to an item (question) correlate well with the responses to other items. For example, suppose we are interested in rating the services of a subway system (based on a scale of 1 to 10) and the sections of the questionnaire include:

1   Respondent characteristics,
2   Fares,
3   Security, and so on.

The section on security has four items (questions), namely:

3.1 Adequacy of the guards,
3.2 Adequacy of lighting,
3.3 Adequacy of cameras, and
3.4 Alertness of the staff.

From the pre-test responses, we correlate the scores for each item with the aggregate scores for all other items. In Table 9.2, the scores for item 3.4 may be correlated with the aggregate scores for all other items.

Standardized Tests

Standardized tests are another way to collect data. They are commonly used in psychological and educational research, such as in an experiment to test mental ability or the effectiveness of a teaching method.

The challenges in designing standardized tests include ensuring that the content is appropriate, there is sufficient time to complete the test, and that the test is not too easy or difficult. Standardization makes such tests less suitable for answering complex questions where there are no simple solutions.

In education testing, it is possible to prepare by teaching to the test, meaning that students learn to ace the test, rather than learn the fundamental principles. What is not tested will be devalued.

Use of Physical Instruments

Physical instruments are widely used in the natural sciences to measure velocity, acceleration, temperature, distance, mass, pressure, weight,

Table 9.2   Item analysis.

| | Item | | | | Sum of scores for 3.1, 3.2, and 3.3 |
	3.1	3.2	3.3	3.4	
John	7	8	7	2	22
Fox	6	8	8	3	22
Joe	7	9	7	5	23
Etc.					

volume, and so on. The decision will depend on factors such as cost, availability, accuracy, precision, ease of use, calibration requirements, and reliability.

Simulation

A simulation model is a mathematical imitation of a real world system. Researchers use simulations to analyze uncertainty, generate statistical distributions, model processes, generate forecasts, or model interactions among variables. It generates its own data, which makes simulation a method of data collection. For example, we may simulate the energy performance in a building using the design drawings and use the results to improve the design before the construction stage.

In business and finance, researchers often use Monte Carlo simulation (MCS) (Laguna and Marklund, 2005; McLeish, 2005; Fontana, 2006) to study the distributions of variables and parameters. In the former case, the basic idea is to first vary the values of different variables such as sales and input costs and then examine how these changes affect profitability. In the case of parameters, we generate artificial data to estimate the values of parameters. For example, suppose we have a regression model such as

$$Y_i = \alpha + \beta X_i + \varepsilon_i$$

where Y_i is the monthly consumption expenditure of the ith household, X_i is the monthly household income, α is the intercept, β is the slope coefficient, and ε_i is the error term. Often, we do not how β varies across different samples. In MCS, we:

- fix the values of α and β, for example, 500 and 0.4 respectively;
- select, say, 100 values of X_i;
- generate 100 values of ε_i using a random number generator;
- compute 100 values of Y_i using the regression equation;
- use the method of least squares to estimate β;
- repeated the process 300 times; and
- plot the 300 estimates of β as a histogram.

The histogram provides a snapshot of how the estimates of β vary.

Economists also routinely use simulation to generate policy forecasts. For example, Kydland and Prescott (1982) used simulations to study the business cycles and draw policy implications. Typically, we use about 90% of the time series data to calibrate the model, leaving the most recent 10% to test its predictions. For example, we may want to test how well a model of interest rates predict future rates. Suppose we have monthly interest rate data from 2000 to 2016. We can use the data from 2000 to 2015 to build the model and then test its predictions against the actual data from 2016.

In agent-based models, researchers want to know how agents interact with their environment (Railsback and Grimm, 2011). For example, Schelling (1978) used agent-based modeling to show how households tend to segregate into racial clusters. He first divided the "urban area" into a grid of cells, one for each household. From an initial random allocation, a household in any cell will survey its eight neighbors to check if they are racially the same or different. If the proportion of different households exceed a certain tolerance (for example, 0.3), the household is "unhappy" and will relocate to another vacant cell. Interestingly, Schelling showed that racial clusters appear after many rounds of shifting.

Although simulation can handle models with many interacting variables, it may degenerate into black box modeling where the results are less certain. The user may not be able to check the computations.

Review of Documents

We may also collect data from published documents such as:

- internal organizational accounting records, sales data, maps, commercial reports, and other miscellaneous records;
- academic journals, directories, magazines, newspapers, commercial reports, reference books, dissertations, theses, encyclopedias, websites, and books; and
- private diaries, letters, memos, photographs, and films.

Permission is required to access these sources. Accuracy is important; if possible, consult the original source that often contains the actual words, intent, methodological details, warnings, and standard errors that are not reported by subsequent users. It is also necessary to verify the authenticity and credibility of the source. For instance, "official" sources may suppress statistics on worksite accidents or use different methodologies or words to make the numbers or organization look good.

References

Blake, R. (1989) Integrating quantitative and qualitative methods in family research. *Families, Systems, and Health,* **7**, 411–427.

Brace, I. (2013) *Questionnaire design.* London: Kogan Page.

Bradburn, N., Sudman, S. and Wansink, B. (2004). *Asking questions: The definitive guide to questionnaire design.* New York: Jossey-Bass.

Fenno, R. (1978) *Home style.* Boston: Little, brown, and Co.

Fontana, M. (2006) Simulation in economics: Evidence on diffusion and communication. *Journal of Artificial Societies and Social Simulation,* **9**, 1–15.

Harris, D. (2014) *The complete guide to writing questionnaires.* New York: I & M Press.

Knoop, T. (2015) *Business cycle economics.* New York: Praeger.

Kydland, F. and Prescott, E. (1982) Time to build and aggregate fluctuations. *Econometrica,* **50**, 1345–1370.

Laguna, M. and Marklund, J. (2005) *Business process modeling, simulation, and design.* New Jersey: Prentice Hall.

Leighton, J. (2017) *Using think-aloud interviews and cognitive labs in educational research.* London: Oxford University Press.

McLeish, D. (2005) *Monte Carlo simulation and finance.* New York: Wiley.

Railsback, S. and Grimm, V. (2011) *Agent-based and individual-based modeling: A practical introduction.* New Jersey: Princeton University Press.

Schelling, T. (1978) *Micromotives and macrobehavior.* New York: W. W. Norton.

Tashakkori, A. and Teddlie, C. (1998) *Mixed methodology: Combining qualitative and quantitative approaches.* Thousand Oaks, California: Sage Publications.

CHAPTER 10

Collection and Processing of Data

Introduction

After developing the research question(s), hypothesis (or framework), research design, and methods of data collection, the next step in the research process is the actual collection and processing of data. The processes are linked, and it is not possible to develop each step independently of prior decisions.

The issues during data collection include:

- access to respondents;
- management of research assistants;
- care and use of equipment;
- protocol for review of documents;
- note-taking;
- establishing the chain of evidence;
- enhancing reliability; and
- tracking of progress.

The issues are similar for all research designs, with minor variations between interpretive and causal studies. These variations will be highlighted below.

Access

Researchers need to gain access to individuals and organizations to gather information. It is the door to field research. Yet, this process of gaining is not well-theorized or documented. The literature is sparse (Feldman *et al.*, 2003).

The *gatekeeper* controls your access to the organization. He is likely to be a leader or senior person in the organization. For example, in a school setting, the gatekeeper is the principal. He will decide whether you can observe and interview the school administrators, teachers, and students. Obviously, you need to address his concerns, such as:

- the purpose of the study;
- why the school has been selected;
- what are you going to do with the results;
- will it disrupt classes; and
- how the school can benefit from participating.

From the last bullet point, we can see that the best way to gain access is probably to show how the other party can benefit from the study. For instance, as an inducement, you may want to share your research findings with the school. This means that the research problem is important to the school. For example, you may want to share whether a new technique for teaching mathematics is effective.

Other ways of gaining access include:

- formal introductions, usually by someone influential such as the chairperson of the school council;
- informal introductions and networks, such as through a teacher or a friend; and
- past links to the school, such as if you are an alumnus.

In a commercial setting, gaining access to workers presents a different set of problems. One issue is the lack of trust between the researcher and the workers. For some workers, the researcher may be a "spy" from management. Further, there may be little incentive for workers to share sensitive views with outsiders.

In the case of questionnaire surveys, access to potential respondents is often through an impersonal mailing list. The researcher has to consider cost (for example, postal, email or web-based), geographical spread, timeliness, ease of obtaining permission, and the likely response rate. The response rate depends on issues such as topic, quality of questionnaire, currency of mailing list, ease of reply, possibility of rewards, and so on.

All forms of access have time limits. It cannot go on forever. For example, if you are interviewing a senior manager, be clear upfront on the expected number of interviews and the period. Do not assume that subsequent access to a respondent to clear up some issues is automatic.

Research Assistants

The management of research assistants includes issues such as planning and budgeting, hiring, training, supervising, scheduling the fieldwork, logistics, safety and health, quality control, and mentoring (Stouthamer-Loeber and van Kammen, 1995).

In the selection of interviewers, honesty, humility, possession of the relevant skills, a good voice, a pleasant personality, good attitude, and a liking for field conditions are pre-requisites. A good command of certain languages is mandatory if not all respondents (for example, tourists) speak the same language. Interviews can be a little frustrating at times, and people with short fuses are unsuitable. It is uncertain whether biases may occur between paid and unpaid field staff, as well as between internal and external field assistants. Motivation and effort are also affected by perceptions of the adequacy of payment.

Training should be provided to field staff members. Someone familiar with the entire research, such as the principal investigator, should conduct the training. The briefing includes the nature and purpose of the study as well as data collection procedures. They should be trained on specific procedures to be followed when contingencies arise, such as (Lavrakas, 1993):

- the respondent is not at home;
- no one is answering the phone;
- the call is directed to an answering machine;
- the call is answered by a non-resident;
- the line has been disconnected;
- the selected respondent is unable to answer because of physical disability;
- there is a language barrier;
- the interview is incomplete; and
- the selected respondent refuses to be interviewed.

A demonstration interview, followed by a trial interview by each field staff member, is recommended. This provides a model for interviewers to follow and an opportunity to correct mistakes. Interviewers may use probes to prompt respondents for their "best guess" answers but they must be mindful of possible bias in the responses. Training should also cover ethical issues and expectations of integrity, such as truthful reporting, processing, and analyses of field data, data protection, avoiding plagiarism, and not infringing copyright.

Scheduling, logistics, safety, and health will need to be considered so that the research assistants are working as a team, work is properly distributed, and the timings and periods of site visits are appropriate and coordinated. Be reasonable in the scheduling; interviewing a large number of people over a short period is not easy, and productivity may fall sharply if motivational aspects are neglected. The researcher is responsible for providing proper clothing, equipment, and safety procedures, such as how to deal with chemicals, radiation, and other hazards.

Finally, a research project provides the opportunity for the researcher to supervise and mentor research assistants. They are not just "a pair of hands" to help you to collect data. In turn, mentored research assistants find the experience more satisfying and productive.

Equipment

Check that equipment such as tape recorders, cameras, and other measuring devices are calibrated and in good working order. Field equipment should be looked after to prevent damage and, for safety reasons, ensure only qualified people handle the equipment. Leaving them unattended invites theft and gives participants the impression of professional irresponsibility.

For interviews, the notebooks, instruction manuals, survey forms, and maps should be properly handled.

Documents

Most research designs require the review of documents for qualitative and quantitative data. There should be a protocol or checklist for the review of such documents to extract the data meaningfully and effectively.

The process begins with an assessment of the types of information required, and hence the types of documents to review. Some of the information may have been published elsewhere, or there are alternate sources of information. For example, information on construction statistics may be published by different government agencies. As far as possible, the researcher should triangulate the data from these sources to minimize errors.

Where possible, use original sources when collecting documentary data to minimize transcription and interpretation errors. The original data may have been reorganized by subsequent users and important footnotes on how these data have been collected may have been omitted.

Note-taking

In many research designs, large quantities of notes need to be managed as a database.

Note-taking requires skill. Observers must be trained on what to observe, how to fill in observation checklists, forms, and log books. Observations should be carried out unobtrusively because subjects may react differently when an observer is busy tracking every step. If the situation is fluid, leave nothing to memory by taking notes as soon as possible. Use color pens to insert footnotes, personal opinion, and follow-up action. It is a good idea to take notes in stages, starting with sketchy notes to keep abreast of what is happening and filling in the details when spare time is available.

As an observer, you may not be sure what is important. Hence, you should jot down the events whenever possible. The perspective may change and what was previously thought to be unimportant may become important.

Chain of Evidence

The notes will have to be organized to trace the chain of evidence. The notes and evidence are usually arranged in temporal sequence or themes. Within each theme, there is still the need to track temporal changes.

In legal terms, a "chain of evidence" is a series of sequential events that account for the actions of a particular person in a specific legal case (for example, a criminal case) from the beginning to the end. The reasoning should be tight for the conclusion to be defensible in court. It is similar to causal mechanism or process tracing (Beach and Pedersen, 2013), and is widely used in forensic science.

Reliability

In interviews, the researcher normally lets the respondent express his views freely, telling or constructing *his* side of the story. Reliability can be enhanced by cross-checking his views with other sources of evidence. For example, if a worker claims that he works long hours, this may be checked with colleagues. Triangulation among observers and other sources of information also minimizes observer bias.

For documentary research, the sources should be reliable and credible, such as data published by reputable researchers and organizations. Triangulation among data sources will also improve reliability.

Tracking of Progress

For interpretive studies, the tracking of research progress is less of a problem once access has been secured and respondents continue to co-operate. This is not so for survey research, where the response rate tends to be more uncertain.

The tracking of research progress also involves field supervision to ensure that research assistants follow field procedures and workloads are reasonable. It is not unusual for supervisors to verify a small portion of the interviews or questionnaires by re-interviewing or asking respondents whether they have been interviewed.

Supervisors should collect survey forms on a regular basis and edit them in the field for legibility and completeness. Where problems occur, these issues are communicated to field assistants and additional training may be necessary. A reminder may be sent, and follow-ups are made soon after the cut-off date.

Data Processing

After data have been collected, the next step is to process them into information suitable for analysis. The processing of qualitative data is part of data analysis rather than a clearly separated process. For quantitative data, processing is necessary to ensure the integrity of the data.

The first stage of data processing is to *edit* the data for errors, contradictions, inconsistencies, and omissions that have escaped preliminary field editing. Falsified data are usually rejected. If errors or missing data have been spotted, then a decision has to be made to discard the information, re-contact the respondent, use an average value among similar respondents, interpolate from other values, or use subject matter knowledge to guess an appropriate value. Care must be exercised in handling outliers just because they do not fit the theory. They may provide a refutation of the theory. Data obtained from published documents may also require editing. They may contain biases such as arbitrary accounting conventions and failure to take into account quality change, price discounts, and reporting errors. They may also contain misprints.

The second stage of data processing is to *transform* the data through conversion, adjustment, or reconstruction. This may involve converting one currency to another, converting monthly to annual income, deriving net from gross values, or rebasing a time series to a new base year.

The third stage is to *code* the data by labeling, classifying, and organizing them for subsequent analysis. Coding is generally straightforward for quantitative data, such as in developing the data table or matrix for regression or multivariate analyses using statistical software. The researcher should avoid heaping, where too much data fall into a particular category. If it occurs, reclassification is necessary.

For qualitative data, coding is the basis for analysis because respondents provide open-ended answers to questions. The development of the *storyline* is fundamental in qualitative studies, so it is not just a matter of labeling and classifying data (Auerbach and Silverstein, 2003). This will be covered in the next chapter.

Big Data

Big data are data sets that are so large that the traditional data processing techniques and software outlined in the previous section are inadequate in dealing with their vastness. The common sources of big data are transactional data and senor data from smart devices (also called Internet of things, or IoT). These include data on financial transactions, transit trips, energy consumption, mobile communications, and so on.

Big data have three characteristics, namely:

- big volume that is based on observations of what happens, rather than merely sampled;

- high velocity, where data are obtained dynamically, such as every second or minute; and
- large variety in terms of text, images, audio, and video, which means the data are also less structured.

Because of these characteristics, traditional multivariate statistical software that run on a single computer will not be able to handle big data within a reasonable amount of time. The storage, maintenance, aggregation, and querying of the database is a major challenge.

Specialized software and hardware, called big data platforms, will have to be used (Grover *et al.*, 2015). The connected servers may play different roles, such as to generate or process data. However, the basic purpose of processing data remains, that is, to detect patterns in data for applications in mass media, transport, manufacturing, finance, cyber security, healthcare, telecommunications, and so on.

References

Auerbach, C. and Silverstein, L. (2003) *Qualitative data: An introduction to coding and analysis*. New York: NYU Press.

Beach, D. and Pedersen, R. (2013) *Process-tracing methods: Foundations and guidelines*. Ann Arbor: University of Michigan Press.

Feldman, M., Bell, G. and Berger, M. (2003) *Gaining access: A practical and theoretical guide for qualitative researchers*. Walnut Creek, California: Altamira Press.

Grover, M., Malaska, T., Seidman, J. and Shapira, G. (2015) *Hadoop application architectures: Designing real-world big data applications*. Sebastopol: O'Reilly Media.

Lavrakas, P. (1993) *Telephone survey methods*. London: Sage.

Stouthamer-Loeber, M. and van Kammen, W. (1995). *Data collection and management*. Thousand Oaks, California: SAGE Publications.

CHAPTER 11

Qualitative Data Analysis

Types of Qualitative Data

Qualitative data, which are used extensively in interpretive and constructivist frameworks (see Chapter 1), comprise texts and visual images. They may be primary data collected from direct observations and interviews in the form of field notes or secondary data drawn from books, diaries, political scripts, articles, newspapers, advertisements, paintings, symbols, artifacts, photographs, films, audio and video recordings, websites, open-ended responses to interviews, recollections, and so on. The analyst should make use of the various types of evidence to paint a coherent picture.

Coding

The raw data require some processing before the actual data analysis. These may include making *back-ups* of original copies, *indexing* sources for easy reference and retrieval, and *transcribing* audio recordings into texts to facilitate analysis. Transcribing does not mean copying verbatim from the recording. The aim is to identify the key points or concepts in the conversation for further analysis.

The *coding* of texts consists of assigning words, phrases, symbols or numbers to each category. The researcher then *annotates* the page margins with informal notes (Table 11.1). The number of codes is a matter of preference, but it should not be too many, resulting in loss of focus.

Codes may be preset or emergent ideas that crop up during coding. These codes may lead to changes in the initial storyline. The researcher may also redefine the codes along the way, such as by merging preset codes in situations where there is no data.

Table 11.1   Example of a coded interview transcript.

Line			Code	Notes
30	Interviewer	How can the park design be improved?		
31	John	We need more cycling paths for cyclists.	04	Cycling
32	Jim	The children's playground is too close to the canal. I have safety concerns.	08	Playground safety
33	Kay	I have safety concerns as well, but for women. The park is poorly lit, with many hiding places.	08	Safety for women
34	Joe	We need shade in our tropical climate.	09	Shade
35	Jane	As our population ages, we need to cater to their needs in terms of access and facilities.	12 13	Access Facilities for older people
36	Joe	Speaking of facilities, there should be sufficient parking spaces for private vehicles.	14	Parking
37	Jim	If you look at East Coast Park, public transport connectivity is poor.	18	Public transport
38	John	Parks should be connected to other green spaces, so that I can jog or cycle for longer distances.	20	Park connectivity
Etc.				

Reflexivity

To be reflexive is to be aware of one's own biases and preconceptions. There are ways for an interpretive researcher to reduce such biases.

First, provide the *background of the researcher* in the final report so that readers are aware of the background and perspective of the researcher. For example, in Peshkin's (1986) study of a fundamentalist Christian school in the US, he provided his background — that of a male Jewish professor of education.

Second, the researcher should provide the *context* or local situation. This requires "thick" rather than "thin" descriptions (Geertz, 1973). A thin description is just a presentation of facts, such as a project manager asking for major changes to the schedule at the meeting. A thick or detailed description provides the context on why major changes to the schedule were required, what were the amendments, the negotiations between the parties, the consultations within each party, the visuals and moods at the meeting, and so on.

Third, the researcher should be aware that his *presence* might affect the study. This is the well-known Hawthorne effect, where the presence of the researchers in an experimental study on factory productivity affected the results. The researchers discovered that changes in factory lighting levels did not have a consistent effect on productivity. When lighting levels were increased, so did productivity. However, when lighting levels were deceased, productivity also went up. In interpretive studies, this effect may also happen, such as if the researcher is taking notes all the time instead of focusing on building trust and observing behaviors and activities in their natural settings.

Fourth, if a researcher's presence can affect the study, then his own thoughts, preconceptions, prior theories, experiences, and feelings can also have a similar effect. Hence, it is important for the researcher to "bracket out" these thoughts, that is, he should put them aside instead of using them to influence the views of the respondents (Moustakas, 1994).

Fifth, it may be helpful to discuss with fellow researchers everyone's personal biases and jot them down in personal journals to create awareness and remind ourselves. It is also useful to report such biases in the research report so that readers are aware of these possible biases and can decide for themselves.

Sixth, as researchers, we should always be mindful that there are two sides to a coin. Each party will tell the story *his* way, such as blaming project delays on other parties or external factors rather than his own mistakes and failings. Other than finger pointing, the reflexive researcher is aware of other reasons or excuses, such as heavy workload, project complexity, regulatory changes, procurement methods, and family issues. We should also be alert to the possibility that one party may *impose* its views of "reality" on another, particularly if the power structure is excessively unbalanced. For example, the school's discipline master may impose his view of what constitutes unacceptable behavior on a student. The student may accept, challenge or negotiate the master's "definition of the situation" (Thomas, 1923). As a second example, a mother and her child may be arguing about what constitutes "good grades" in school.

Seventh, we can also use *crosschecking* or triangulation to reduce researcher bias. For instance, if a person claims that he is the unofficial leader of the team, the researcher can check it with other team members or another researcher.

Finally, we need to pay attention to the use of language or *discourses*. Words are not neutral; speakers deliberately choose certain words to persuade. A related term is ideology, such as the ideology of a free market. It is a coherent set of beliefs by supporters of the free market, such as political and economic freedom, efficiency, government failures, and so on. Paradoxically, self-regulating "free" markets are rare; most markets are regulated.

In summary, reflexivity is about reducing researcher bias through continuous self-questioning, not taking anything for granted — not even language. There is awareness of the presence of self-bias, and recognition of multiple views, explanations, and options. It is a key principle of qualitative research.

We will discuss the various ways of analyzing qualitative data below. The list is not exhaustive, and readers should consult books on qualitative research for details.

Narrative

The narrative is a non-fiction storyline that guides the entire data analysis (Riessman, 2007; Clandinin, 2013). It is an intelligible story of human actions and the resulting events in temporal order. The framework or hypothesis provides the guideposts for the story. The researcher then proceeds to code the data, knowing that the narrative may change as the research proceeds to discover new ideas and evidence. The narrative, then, is not a mindless articulation of curious historical facts.

Stories may have structures or *plots*, such as a difficult start, a mid-story crisis, a turning point (epiphany), and a happy resolution. These plots contain *conflicts* or struggles in particular *settings* or environments. Stories of nation building in historical case studies may be of this genre (White, 1990), and they typically include a "hero" (*protagonist*) who organized the struggle against colonial or other forms of oppression (*antagonist*). Sometimes, the structure is reversed: a happy start, a mid-story crisis, a turning point, the current unhappy state, and the economy is in a mess. The struggles may also be internal to the actor, rather than against an external protagonist. Finally, stories take particular *points of view* or theoretical perspectives. For example, a researcher may use dependency theory (Amin, 1976) to write the story of a country's struggle against colonialism.

The researcher adopts a particular *style* or language to tell the story. For example, metaphors are widely used to persuade readers — such as in this example: "The public and private sectors entered into a *partnership* to develop the project." This conveys the impression of a co-operative win-win arrangement to the public.

The term "analytic narrative" refers to the attempt to construct general theories out of individual narratives or case studies and then subject the explanation to empirical tests (Bates *et al.*, 1998).

Discourse Analysis

The purpose of discourse analysis is to *deconstruct* texts to reveal the ways they create certain views of "reality" and sustain ways of life through such cognition and power relations (Derrida, 1976; Foucault, 1991). Persuasion occurs through *appeals* to reason, statistics, partial evidence, tradition, values, expertise, simulations, word choice, analogies, metaphors, and anecdotes that avoid the burden of proof.

For example, consider how a government may use discourse (as shown in italics below) to argue its case for shrinking government expenditure (Seymour, 2014). Here, the *culprit* is *overspending*. The former UK Prime Minister Margaret Thatcher likened the government to a small corner shop or household that needs to keep its finances in order. We may use discourse analysis to expose the weaknesses of this *analogy*. First, unlike the government, a household cannot print money, so the two entities are not comparable. Second, Keynes (1936) identified the paradox of thrift as a problem for the economy. If households save too much, consumption demand will fall and businesses will not invest if the future is bleak. Hence, what is good for the household in balancing its budget may not be good for the country.

Discourse is seldom something we imagine or cook up to fool people. It is real and material, for overspending is evident in the official *statistics* on the government's debt. The widening government deficit then leads to a fiscal *crisis*, because governments cannot *pile* debt upon debt as if it is business as usual. We, as a *nation*, *owe* it to future generations to bring the *unsustainable* debt down, so that business confidence will grow and firms will invest in the future.

Why do some governments overspend? Mass and social media have plenty of theories. Some people attack *irresponsible* politicians who promise a lot but deliver little. Others blame *self-serving* bureaucrats who do not act in the public interest. A third group of *experts* focuses on policy failures, such as the strategy of import substitution industrialization. The structuralists blame deindustrialization (Bluestone and Harrison, 1982); as manufacturing firms go offshore in search of lower production costs, the rise in structural unemployment necessitates additional *welfare* spending.

In contrast, those on the far Right such as Mrs. Thatcher and former US President Ronald Reagan attacked welfare and governance, in particular:

- inefficient civil servants;
- too many services for the poor;
- generous welfare benefits that create the disincentive to find work and demean the *value* of hard work; and
- lack of transparency in awarding contracts and licenses.

Tired of frequent construction cost over-runs in public projects, Mrs. Thatcher introduced *partnerships* with the private sector, called Public Finance Initiatives (now more commonly known as Public Private Partnerships). She portrayed the British labor unions as *uncooperative*, *self-interested*, and *shortsighted*, making it difficult for firms to adjust to changing cyclical and structural conditions.

How can we address these government *failures*? The neoliberal free market discourse works in reverse, starting from political reform so that we have better *leaders*. Next, these leaders must have the political will to balance the budget and reduce government debt. This will lower interest rates by reducing the demand for government borrowings. Business confidence will improve, and the economy will grow again. There is a long list of ways to reduce the fiscal deficit and impose market discipline, such as:

- selling assets through privatization;
- reducing public subsidy, including essential commodities;
- the adoption of market-based performance measures, and hiring and promoting on merit;
- more stringent *workfare* benefits to encourage positive work attitudes;
- reforming and deregulating the goods, land, financial, and labor markets to make them competitive and flexible; and
- implementing good governance to make governments accountable to the public.

Obviously, there will be winners and losers when governments implement certain policies. For example, the World Bank and International Monetary Fund (IMF) implemented a series of free market structural adjustment programs (SAPs) in the 1970s and 1980s as a condition for loans to governments who ran out of money during the debt crises. These programs have been known to hurt the poor (Lundberg and Squire, 2003; Hertz, 2004). Since the mid-1990s, poverty reduction has been included as a goal in SAPs.

In summary, people deliberately use their words carefully to convey certain impressions to persuade others. Deconstruction is a method of analyzing discourses to expose such acts of persuasion.

Content Analysis

Content analysis involves *quantifying* the contents of the written or digital texts, such as looking for the occurrences of particular words or images. For example, we can count the number of times politicians mention a particular development project in the news. In qualitative applications of content analysis, researchers may not use pre-conceived categories in the hope of finding emergent categories.

In McClelland's (1962) study on the *achievement motive* and economic growth, he argued that one could not simply infer real motives from what people said. Hence, McClelland tried to measure the achievement motive indirectly by examining the content of children's books for stories. He differentiated stories that stressed success from those that stressed failures and misery. He then correlated the frequency counts with economic growth and found that countries with children's books that stressed success tended to experience higher economic growth compared to those that did not.

However, critics argued that McClelland did not establish the correlation because the economic growth data, from the 1920s to the late 1950s, were not reliable. Several studies using updated data found little correlation (Mazur and Rosa, 1977; Gilleard, 1989). Further, the values in children's stories may merely reflect those of the authors rather than those taught to children at home or in school. Finally, to focus on a single variable, achieving motivation, as the cause of economic growth is naïve. Current work emphasizes the political-economic system, that is, the State and the free market, rather than a small aspect of culture. For example, even if it is true, it begs the question why some societies are not motivated to achieve. A farmer will not plant new crops if they are going to be confiscated, heavily taxed, or stolen by harvest time. This raises the issues of property rights, taxation, and law and order, which are institutions of the production system.

Grounded Theory

The aim of grounded theory is not so much to analyze content as to *generate theory* from a systematic analysis of qualitative data (Glaser and Strauss, 1967). The approach is inductive, from evidence to the discovery of theory. Hence, grounded theory is a form of qualitative data analysis to generate theory. It is, by itself, not a theory.

In the first stage, the researcher reviews the sparse literature to develop a *preliminary framework* to guide data collection. The research design is usually a case study using in-depth probes to interview selected respondents who can best help to develop the theory. A survey may also be used but it will not be a mass survey. Rather, the purpose of *theoretical*

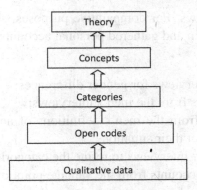

Fig. 11.1   Steps in grounded theory.

sampling is to interview more respondents intensively to discover emergent ideas to develop a theory. The coding starts with open codes and progressively narrows through *constant comparisons* between the existing codes and new data to build up a small set of categories (axial coding) and concepts to develop the theory (Fig. 11.1).

What, then, is the sample size in theoretical sampling? The sampling continues until the point of *theoretical saturation* where there are diminishing returns from examining more cases. Finally, the researcher connects the fragmented ideas into a coherent theory (Glaser, 1978). This conceptual leap is not easy.

As an example of the grounded theory approach, we consider Charmaz's (1994) study on the identity dilemmas of chronically ill men. It is a psychological phenomenological study of the lived experiences of participants. According Charmaz, the literature does not explicitly address this issue of the identity dilemmas of such men. However, there is a prior framework to guide the study, namely, the *phases* of:

- awakening to the illness;
- accommodating the uncertainty of the effects of the illness;
- defining illness and disability; and
- preserving self amid loss and the accompanying change.

Charmaz used theoretical sampling to interview 20 men who have chronic illnesses, and seven were interviewed more than once, resulting in

40 in-depth interviews. For comparative purposes, she also interviewed chronically ill women and gathered personal accounts from both men and women. She then:

- examined the interviews for gender differences;
- developed themes from the men's interviews;
- built categories from the men's definitions of and taken-for-granted assumptions about their situations;
- conducted further interviews to refine the categories;
- reread personal accounts from the vantage point of gender issues;
- studied a new set of personal accounts; and
- made comparisons with women on selected key points.

I will not discuss all the points here, particularly comparisons with women.

The raw data were first transcribed. Table 11.2 is a modified version of the first phase when the illness strikes (i.e. awakening to illness) and it contains added codes to indicate how a theme (for example, vulnerability or envy) appears.

Charmaz then continued to transcribe and categorize the data for other phases, which I will not discuss here. In the end, the paper was exploratory and did not explicitly generate any theory. This is not surprising because making the conceptual leap is difficult when the theory is still not well-developed. Nonetheless, the study raises the question on the factors that shape whether a man with a chronic illness will reconstruct a positive identity or sink into depression. Charmaz pointed out a two-edged sword: men who value autonomy and personal power in the face of danger tend to take risks to be active again, but the inability to do so may sink them into depression. Further, if they feel that no options are available, they are more likely to slip into depression.

Lastly, there is the issue of rigor in the grounded theory approach. Clearly, it is inappropriate to apply the criterion of testability or predictability used to evaluate causal studies. However, are the interpretations reasonable? Charmaz wanted to probe into the identity dilemmas of chronically ill men from *their* perspectives, which are not directly observable. However, we can still evaluate her methodology and how she analyzed the data.

Table 11.2 Example of an interview transcript.

Line			Code	Notes
29		John, competitive cyclist, hard-driving no-nonsense businessman, former Vietnam War veteran, breadwinner, and head of household, all masculine identities. Business failed just before he had a heart attack, forcing his wife to work.		
30	Interviewer	How did you come to identify yourself?		
31	John	I didn't know who I was for a while ... all of a sudden, the hardest thing to accept is, "Hey, you are vulnerable."	04	Vulnerability
...				
90	Interviewer	How did you come to identify yourself?		
91	David	It's a mid-life crisis ... I am too young to die ... But that's life.	04	Vulnerability
92		The future is bleak ...	05	Bleak future
93		I envy my mates who are health y...	06	Envy
94		I am also depressed ...	07	Depression
95		But my wife and family have been very supportive ...	08	Family support; he cried
...				

References

Amin, S. (1976) *Unequal development: An essay on the social formations of peripheral capitalism*. Los Angeles: University of California Press.

Bates, R., Grief, A., Levi, M., Rosenthal, J. and Weingast, B. (1998) *Analytic narratives*. New Jersey: Princeton University Press.

Bluestone, B. and Harrison, B. (1982) *The deindustrialization of America*. New York: Basic Books.

Charmaz, K. (1994) Identity dilemmas of chronically ill men. *The Sociological Quarterly*, **35**(2), 269–88.

Clandinin, J. (2013) *Engaging in narrative inquiry*. London: Routledge.

Derrida, J. (1976) *Of grammatology*. Baltimore: Johns Hopkins University Press.

Foucault, M. (1991) *Discipline and punish: The birth of a prison*. London: Penguin.

Geertz, C. (1973) *The interpretation of cultures*. New York: Basic Books.

Gilleard, C. (1989) The achieving society revisited. *Journal of Economic Psychology*, **10**(1), 21–34.

Glaser, B. and Strauss, A. (1967) *The discovery of grounded theory*. Chicago: Aldine.

Glaser, B. (1978) *Theoretical sensitivity*. Mill Valley, California: Sociological Press.

Hertz, N. (2004) *The debt threat*. New York: Harper Collins.

Keynes, J. (1936) *The general theory of employment, interest and money*. London: Macmillan.

Lundberg, M. and Squire, L. (2003) The simultaneous evolution of growth and inequality. *Economic Journal*, **113**, 326–344.

Mazur, A. and Rosa, E. (1977) An empirical test of McClelland's "achieving society" theory. *Social Forces*, **55**(3), 769–774.

McClelland, D. (1962) Business drive and national development. *Harvard Business Review*, **40**(4), 99–113.

Moustakas, C. (1994) *Phenomenological research methods*. London: Sage Publications.

Peshkin, A. (1986) *God's choice*. Chicago: University of Chicago Press.

Riessman, C. (2007) *Narrative methods for the human sciences*. Thousand Oaks: SAGE.

Seymour, R. (2014) *Against austerity*. London: Pluto Press.

Thomas, W. (1923) *The unadjusted girl*. Boston: Little, Brown, & Co.

White, H. (1990) *The content of the form: Narrative discourse and historical representation*. Baltimore: Johns Hopkins University Press.

CHAPTER 12
Quantitative Data Analysis I: Survey Data

Nature of Survey Data

Survey data usually contain large amounts of information on respondent characteristics and their views. The data may be spatial, cross-sectional, or temporal, and include frequency counts, ratings, ranks, and continuous variables.

The measurement errors from survey data vary. These errors may be large if we are gathering sensitive information such as income, wealth, religion, race, political opinions, monetary contribution, and taxation. For example, respondents may not be truthful in revealing their demand for a public good if they are required to contribute towards its provision based on their responses (Clarke, 1971). There is an incentive to "free ride" by not paying one's fair share and letting other people foot the bill.

In this chapter, we will consider the use of simple data analytic techniques involving means, variances, ranks, indexes, frequencies, and possible relations among the variables.

Exploratory Data Analysis

The purpose of exploratory data analysis (EDA) is to examine simple data patterns such as:

- relations among variables;
- the presence of outliers;
- trends and turning points; and
- distributional assumptions (Tukey, 1977).

The display or presentation of data may take the form of simple tables, texts, plots, graphs, and charts. EDA is usually the first step in quantitative data analysis where we explore patterns in data prior to more rigorous statistical analyses.

The correlation between any *two* continuous variables may be visually inspected by plotting the data in a scatter diagram. The correlation, if it exists, may be linear or nonlinear. If it is linear, the data pattern looks like a line. If it is nonlinear, the data pattern resembles a curve. Such inspections are simple to carry out, and are useful in deciding whether to include a variable or adopt a linear or nonlinear functional form for the regression. It is worth repeating that correlation is not causation. Further, even if the correlation is close to zero, the two variables may still be nonlinearly related. For example, a V-shaped data pattern may have a zero linear correlation coefficient but the variables are nonlinearly related (Fig. 12.1).

Outliers are extreme points that are "far" from the main cluster of points, such as the point (9, 3) in Fig. 12.1. If they are not measurement errors, outliers that result from variability are important because they may provide a refutation of the theory. A statistical distribution with more outliers or extreme values has a heavy or fat tail. In contrast, a normal distribution has a thin tail with fewer extreme values. An example of a non-normal distribution is stock prices, which has more outliers or "black swan" events (Taleb, 2010).

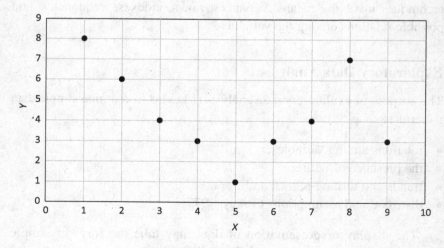

Fig. 12.1 Nonlinear correlation.

For time series data, we can plot them to see if they trend together, move in opposite directions, or move randomly. For each series, it may be possible to determine the amplitudes (magnitudes of fluctuations) and turning points, and hence the period of the cycle, that is, the mean length of time measured from peak to peak, or from trough to trough. For economic data, these cycles may be short (seasonal), mid-length (5 to 10-year business cycles), or long (50-year waves). We can use spectral analysis to detect fluctuations such as cycles or less regular waves (Stoica and Moses, 1997).

For frequency data, we use histograms to check the distribution, such as its shape, mean, and variance. Such inspections are often necessary because many statistical tests contain distributional assumptions such as normality.

Spatial Data

There are many ways of analyzing spatial data such as:

- the use of a road *network* to study its complexity or find the shortest or least-cost path;
- examining *spatial point patterns* to identify spatial clusters, such as the outbreak of diseases in a city;
- examining *spatial correlations* among neighboring points because the distribution of spatial objects need not be random;
- *interpolating* spatial data to draw contours in terrain mapping or to predict the value of a function (for example, mineral deposits) from neighboring points;
- *adjusting* measured spatial coordinates in engineering survey networks for internal consistency (Schofield and Breach, 2007); and
- solving *inverse problems* (Hansen, 2010), for example, the use of surface gravity measurements to predict the extent of mineral deposits below the earth's surface.

Readers who are interested in these techniques may consult Cressie (1993), Lloyd (2010), and Chang (2016). As an example, we consider the data for three weather stations (A, B, and C) with given coordinates (x, y) and our task is to estimate the value of rainfall (z) at point D (Fig. 12.2).

Table 12.1   Data for interpolation example.

Station/Point	x	y	z
A	1	1	0
B	2	4	5
C	5	3	10
D	2	3	?

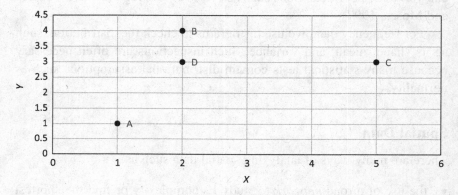

Fig. 12.2   Spatial representation of rainfall data.

We can interpolate the data in many ways, such as through the inverse distance weighted method (IDW), kriging, or trend surface analysis (Stein, 1999). I shall illustrate the inverse distance IDW method, which gives greater weight to nearby points through an inverse square distance weighting function. The predicted rainfall value at D is the weighted sum of the known values at other stations, that is,

$$z_D = \Sigma \lambda_i z_i.$$

Here λ_i are the weights, that is,

$$\lambda_i = (1/d_i^2)/(\Sigma \, 1/d_i^2).$$

The ith subscript refers to the station or point. The first step is to compute the distances (d_i) between each station and D (Table 12.2).

Next, we compute the denominator in the weights:

$$\Sigma \, 1/d_i^2 = 1/(\sqrt{5})^2 + 1/1^2 + 1/3^2 = 1.31.$$

Table 12.2   Computed distances.

From	To	Distance, d_i
A	D	$\sqrt{5}$
B	D	1
C	D	3

Hence,

$$\lambda_A = (1/5)/1.31 = 0.15;$$
$$\lambda_B = (1/1)/1.31 = 0.76; \text{ and}$$
$$\lambda_C = (1/9)/1.31 = 0.08.$$

The predicted rainfall value for D is

$$z_D = \Sigma \, \lambda_i z_i = 0.15(0) + 0.76(5) + 0.08(10) = 4.60.$$

Index Numbers

We use index numbers to track temporal changes in prices, costs, water quality, and so on. To construct a *nominal* price index, we use market prices. To convert a nominal price index into a *real* price index, we need to deflate it by a suitable inflation index such as the consumer price index (CPI). There are other types of deflators such as the Gross Domestic Product (GDP) deflator and producer price deflator.

Example

If the nominal house price index is 105 and the CPI is 103, then the real house price index is 105/1.03 = 101.9.

In the discussion below, we will be constructing nominal price indexes. You may convert them into *real* price indexes by using a suitable deflator.

Price index for homogeneous product

To construct the price index of a relatively homogeneous product such as an apple, we merely need to decide on the "standard" apple (for example, rose apple) to remove quality differences among apples. Then if $P_0, P_1, \ldots,$ P_n are annual prices, we divide each price by its base period price to obtain the following index:

$$1, P_1/P_0, \ldots, P_n/P_0.$$

The index is 1 in the base period because $P_0/P_0 = 1$. We often implicitly multiply the index by 100 to shift the decimal point (for example, $1.031 \times 100 = 103.1$).

For example, if the annual apple prices are 50c, 55c, 60c, and 80c respectively, then the price index is

$$1, 55/50, 60/50, 80/50,$$

that is,

$$100, 110, 120, 160.$$

Price index for heterogeneous product

In practice, we often need to construct price indexes for heterogeneous products such as houses or computers. There are many types of housing, such as public flats, private apartments, condominiums, and landed houses. Further, the houses are distributed spatially, and location has a major impact on house prices.

Suppose we wish to construct the quarterly urban house price index for a region or province with two cities, A and B. The cities have different populations, say, 3 million and 1 million respectively, in the base period. Each city is divided into five residential sectors (Central, North, South, East, and West). The house price index will need to use prices from both cities, weighted according to population.

Let us first consider how we may construct a price index for public flats in city A. For each quarter, we can use public housing resale data from each sector. A common choice is to use the median sales price because it is less sensitive to extreme values than the mean sales price. From the five median sales prices, one from each sector, we can find the average value using the sales volume in each sector as weights, that is,

$$P = w_1 P_C + \cdots + w_5 P_W.$$

Table 12.3 Price index and sales volume by type of property in city A for Q1, Year X.

Property type	Price index	Sales volume
Public flats	120	2,000
Private apartments	130	3,000
Landed houses	150	1,000

Here P is the weighted price for resale public housing units, P_c is the median house price for the Central sector, and P_w is the median house price for the West sector. The weight for each sector (w_i) is its housing sales volume divided by the total sales volume from all five sectors for the quarter.

We may construct the price index for other types of houses in a similar manner, that is, by using the weighted median prices. From the price indexes for the different types of houses for a particular quarter in Year X, we can then construct the aggregate house price index for city A using the sales volume as weights (Table 12.3).

The final step is to combine the price indexes from the two cities using city population or housing transactions as weights. Because we use the same base period weights in each quarter, it is a *Laspeyres Index*. Over time, the city populations will change and we need to update the weights, say after every 10 years. This is called rebasing an index.

Accounting for quality change

Ideally, we want the price index to track pure price changes, that is, how the price of a constant-quality public flat changes over time. The use of median prices above does not satisfy this requirement.

To control for quality changes, we need to specify a "standard flat" (for example, a standard 4-room public flat) and track how prices change over time. However, a weakness of this approach is that the characteristics of the "standard flat" will also change over time. In other words, it is not of constant quality.

To overcome this problem, we may use the hedonic price method discussed in Chapter 8. Here, we take a sample of public flats and regress the price on housing characteristics to discover implicit prices for each period. Recall that the estimated regression coefficients are the implicit prices. We then use these coefficients to predict the price of a "standard" flat with a pre-determined bundle of characteristics. Finally, we use these prices to construct the price index by comparing them with the base period price (Palmquist, 1984).

The hedonic price model requires large amounts of data. A more parsimonious approach is to use repeat sales data, on the assumption that house quality does not change between sales (Bailey *et al.*, 1963; Goetzmann, 1992). Suppose

$$P_i = \beta_1 + \beta_2 X_2 + \cdots + \beta_k X_k + \lambda_i T_i + u.$$

Here the βs are parameters (coefficients), Xs are housing characteristics, T_i is the ith period when the house is sold, and u is the error term. If the same house is subsequently sold in the jth period,

$$P_j = \beta_1 + \beta_2 X_2 + \cdots + \beta_k X_k + \lambda_j T_j + v$$

where v is another error term. The price difference is

$$P_j - P_i = \Delta P = \lambda_j T_j - \lambda_i T_i + w$$

where $w = v - u$ is another error term. More generally, if T is the period in which the house is sold, then

$$\Delta P = \lambda_1 T_1 + \ldots + \lambda_h T_h + w.$$

Here $T_i = -1$ if the house is first sold in that time period, and $+1$ if it is sold in the second time period. Note that the repeat sales model does not require data on any of the housing characteristics (Xs).

To convert the estimated λs from the regression into a house price index, we redefine ΔP as $\log(P_j/P_i)$ rather than just a difference in prices, that is,

$$\log(P_j/P_i) = \lambda_1 T_1 + \cdots + \lambda_h T_h + w. \tag{12.1}$$

As before, log(.) is the natural logarithm. Because we want our price index $I_1, I_2, \ldots, I_h$ to reflect that

$$P_j/P_i = I_j/I_i,$$

we obtain

$$\log(P_j/P_i) = \log(I_j/I_i) = \log(I_j) - \log(I_i).$$

Comparing this equation with Equation (12.1) gives

$$\log(I_1) = \lambda_1;$$
$$\log(I_2) = \lambda_2;$$
$$\text{etc.}$$

Thus,

$$I_1 = \exp(\lambda_1);$$
$$I_2 = \exp(\lambda_2);$$
$$\text{etc.}$$

where exp(.) is the exponential function. This gives the desired house price index.

Our final example on index number is the productivity index. Consider a production function $f(.)$ where output (Q) is a function of capital input (K), including land, and labor input (L). Then

$$Q = f(K, L).$$

For estimating purposes, it is usual to specify a Cobb-Douglas production function, that is, we assume that

$$Q = AK^{\alpha}L^{1-\alpha}.$$

Here A represents the efficiency in which an economy, industry or firm uses the inputs to produce the output. It is called the Total Factor Productivity (TFP). It is "Total" because it estimates productivity from capital and labor inputs. A partial productivity measure such as labor productivity holds capital input constant. For given quantities of K and L, an economy, industry or firm with larger values of A will produce more output. TFP reflects factors other than inputs of capital and labor that affect output. These factors include better management and organization, learning economies, scale economies, worker incentives and effort, research and development, the regulatory environment, and so on. The large number of variables that affects TFP is the main reason why it is hard to pin down the causes of productivity changes, and consequently, why it is not easy to improve productivity.

Differentiating the Cobb-Douglas function with respect to time (t) and using the basic calculus result that

$$d[\log(x)]/dt = (dx/dt)/x,$$

we have,

$$A^* = Y^* - \alpha K^* - (1 - \alpha)L^*,$$

where A^* is the rate of productivity growth, Y^* is the rate of output growth, K^* is the capital growth rate, and L^* is the labor growth rate. Observe that if $K^* = L^* = 0$, then A^* may also be interpreted as the rate of output growth if inputs do not grow.

The parameter α represents capital's share of the output, which is about 0.35 for many economies (Weil, 2009). As an example, if for a particular year $Y^* = 3\%$, $K^* = 2\%$, and $L^* = 1\%$, then the rate of productivity growth is

$$A^* = 3 - 0.35(2) - 0.65(1) = 1.65\%.$$

We can use the above formula to estimate productivity growth for an industry using suitable values for α. For more capital-intensive industries, the value for α will be greater than 0.35 to reflect the greater share of capital's contribution to output.

Ratings

Many surveys ask respondents to rate k factors that affect some event, process, or activity in terms of their frequency of occurrence (for example, 1 = Rare, 5 = Often) and criticality (for example, 1 = Not critical, 5 = Critical).

In Table 12.4, the "Count" column refers to the number of responses for each question. For instance, 100 respondents answered the question on Factor 1, but only 90 respondents replied to the question on Factor 2. The next two columns contain the mean ratings for frequency of occurrence and criticality respectively. For instance, an event may occur frequently (for example, rain) but its effects are not critical. On the other hand, an event may occur infrequently (for example, a site accident) but its effects are

Table 12.4   A rating table.

Factor	Count	Mean frequency rating (a)	Mean criticality rating (b)	Impact [(a) × (b)]
1	100	3.4	4.5	15.3
2	90	2.0	3.2	6.4
...		...	...	...
k	70	3.0	2.5	7.5

Table 12.5   Selection of contractor using judgmental weights.

		Bidder			Bidder		
		X	Y	Z	X	Y	Z
Criteria	Weight	Original score			Weighted score		
A	0.2	6	7	8	1.2	1.4	1.6
B	0.1	5	6	9	0.5	0.6	0.9
C	0.1	9	6	5	0.9	0.6	0.5
D	0.4	6	8	7	2.4	3.2	2.8
E	0.2	6	8	5	1.2	1.6	1.0
Total	1.0				6.2	7.4	6.8
Bid ($m)					100	105	108
Value ratio					0.062	0.070	0.063

critical. Finally, the impact of each factor is then computed in the last column. Factors with high impacts require close attention and monitoring.

We may apply judgmental weights to the ratings. For instance, bidders for a project may be selected using suitable weights on items such as track record and expertise (A), financial strength (B), workload (C), bid price (D), and schedule (E) (Table 12.5). We multiply the original scores by the weights to obtain the weighted scores. The total weighted score for each bidder is then divided by the bid price to obtain the value ratios. In this case, bidder Y provides the best value.

Alternatively, the weights may be derived using pairwise comparisons. In Table 12.6, a judging panel makes pairwise comparisons between

Table 12.6 Derivation of weights.

Criteria	A	B	C	D	E	Frequency	Weight
A		A	C	D	A	2	0.2
B	A		B	D	E	1	0.1
C	C	B		D	E	1	0.1
D	D	D	D		D	4	0.4
E	A	E	E	D		2	0.2
Total						10	1.0

pairs of criteria. In the first row, criterion A is considered to be more important than B, less important than C and D, and more important than E. We then transpose the first row results to the first column by symmetry (shown as boldface letters).

In the second row, B is considered to be more important than C but less important than D or E. Again, we transpose the results to the second column (shown as underlined letters). For each row, we note the frequency of the criterion from which weights are derived. For instance, in the first row, criterion A appears twice. In the second row, B appears once, and so on. A criterion with zero frequency is dropped from consideration.

Ranks

In survey research, respondents may be asked to rank their preferences on projects, products, designs, and so on. In Table 12.7, five respondents have been asked to rank their preferences for three projects A, B, and C. For example, Fox prefers C to A to B. Under the null hypothesis of no difference in preferences, the *Friedman test* statistic for large n is given by

$$F_r = -3n(k + 1) + hS$$

where
n = number of respondents = 5;
k = number of categories = 3 (that is, A, B, and C);
$h = 12/[nk(k + 1)] = 12/60 = 0.2$; and
$S = 8^2 + 11^2 + 11^2$ = sum of squares of column ranks = 306.

Table 12.7   Rankings of three projects.

Respondent	A	B	C
Fox	2	3	1
Joe	1	2	3
John	1	3	2
Ken	3	1	2
Ben	1	2	3
Total	$R_1 = 8$	$R_2 = 11$	$R_3 = 11$

Table 12.8   A 3 × 3 contingency table.

	Profitability			
Firm size	L	M	H	Total
S	60	20	10	90
M	20	30	10	60
L	10	20	10	40
Total	90	70	30	$N = 200$

It can be shown that F_r is distributed as Chi-square with $k - 1$ degrees of freedom. Hence,

$$F_r = -60 + 0.2(306) = 1.2.$$

The 0.05 critical value for a Chi-square variable with 2 degrees of freedom is 5.99. We do not reject the null hypothesis. For further details on how to model and conduct rank tests, see Marden (1996).

Contingency Table

A contingency table tests whether two nominal variables are independent. For example, from a survey of 200 construction firms, we wish to test if firm size (classified as Small, Medium, or Large) is related to profitability (Low, Medium, or High).

The *observed* frequencies are shown in Table 12.8. The research hypothesis is that larger firms, with better access to markets and inputs, are likely to be more profitable than smaller ones. However, they may be less dynamic and flexible. Obviously, good definitions of firm size and profitability matter; otherwise, the frequency counts will not be meaningful.

There are r row and c column categories, so this is an $r \times c$ contingency table. Here $r = c = 3$. If firm size and profitability are independent (H_0), we can disregard the breakdown of the cell frequencies and compute the *expected* frequencies as the product of the marginal totals divided by the grand total (N):

40.5	31.5	13.5
27	21	9
18	14	6

For example, for the first row, $90(90)/200 = 40.5$, $70(90)/200 = 31.5$, and $30(90)/200 = 13.5$. The test statistic is the sum of the normalized squares of the difference between observed and expected frequencies:

$$Q = (60 - 40.5)^2/40.5 + (20 - 31.5)^2/31.5 + \cdots + (10 - 6)^2/6 = 28.28.$$

It can be shown that Q is distributed as Chi-square with $(r - 1) \times (c - 1)$ degrees of freedom. From Appendix 1, the 0.05 critical value for 4 degrees of freedom is 9.49. We reject H_0, and conclude that firm size does matter when it comes to profitability. For details on contingency tables, see Kateri (2016).

References

Bailey, M., Muth, R. and Nourse, H. (1963) A regression method for real estate price index construction. *Journal of the American Statistical Association*, **58**, 933–942.

Chang, K. T. (2016) *Introduction to geographic information systems*. New York: McGraw-Hill.

Clarke, E. (1971) Multipart pricing of public goods. *Public Choice*, **11**, 17–33.

Cressie, N. (1993) *Statistics for spatial data*. New York: Wiley.

Goetzmann, W. (1992) The accuracy of real estate indexes: Repeat sales estimators. *Journal of Real Estate Finance and Economics*, **5**, 5–53.

Hansen, P. (2010) *Discrete inverse problems*. Philadelphia: SIAM.

Kateri, M. (2016) *Contingency table analysis*. New York: Birkhauser.

Lloyd, C. (2010) *Spatial data analysis*. London: Oxford University Press.

Marden, J. (1996) *Analyzing and modeling rank data*. London: Chapman and Hall.

Palmquist, R. (1984) Estimating the demand for the characteristics of housing. *Review of Economics and Statistics*, **66**, 394–404.

Schofield, W. and Breach, M. (2007) *Engineering surveying*. London: Butterworth-Heinemann.

Stein, M. (1999) *Interpolation of spatial data*. Berlin: Springer.

Stoica, P. and Moses, R. (1997) *Introduction to spectral analysis*. New Jersey: Prentice Hall.

Taleb, N. (2010) *The black swan*. New York: Random House.

Tukey, J. (1977) *Exploratory data analysis*. London: Pearson.

Weil, D. (2009) *Economic growth*. New York: Pearson.

CHAPTER 13

Quantitative Data Analysis II: Experimental Data

Parallel Group Design

Unpaired t test

We use the unpaired *t* test for two independent samples to analyze data for the classical experimental design and parallel group design.

If the experimental results for each group do not affect one another, we say that the groups are statistically independent. We further assume that the samples come from a normal population with mean μ and variance σ^2. The samples should be of similar sizes (n_1 and n_2 respectively), and not too small so that the sample variances can be reasonably estimated. For sample sizes greater than 30, the *t* and normal distributions are identical.

Under the null hypothesis (H_0) that there is no difference in mean scores between the two groups, the test statistic is

$$t = D/S.$$

It follows the *t* distribution with $n_1 + n_2 - 2$ degrees of freedom. Here *D* is the difference in sample means and *S*, the standard deviation of *D*, is computed from

$$S^2 = V/n_1 + V/n_2$$

where *V* is the pooled variance, our estimate of σ^2. We compute *V* by using the weighted average of the sample variances s_1^2 and s_2^2, that is,

$$V = [(n_1 - 1)s_1^2 + (n_2 - 1)s_2^2]/(n_1 + n_2 - 2).$$

Example

A project company recruited 20 trainees and put one group of 10 trainees (Group 1) through a new training program. The other group followed the usual training program. The trainees are then assessed on their performance on a 1 to 10 scale, from poor to excellent. The sample data are shown below:

Group	Mean	Variance	Size
1	7	1.0	10
2	6	2.0	10

Here $D = 7 - 6 = 1$, and

$$V = [9(1.0) + 9(2.0)]/18 = 1.5.$$

Thus,

$$S^2 = 1.5/10 + 1.5/10 = 0.3.$$

Hence,

$$t = D/S = 1/\sqrt{0.3} = 1.825.$$

From statistical tables, the 0.025 critical value for t_{18} is 2.101. We do not reject the null hypothesis. Statistically, there is no difference in performance between the groups.

Test of proportions

We may extend the unpaired t test to ascertain if there is a difference in proportions between two large independent samples, that is, we test

$$H_0: \pi_1 = \pi_2 \text{ against } H_1: \pi_1 \neq \pi_2$$

where π_i, $i = 1, 2$ are the population proportions. Under H_0, the test statistic

$$Z = (p_1 - p_2)/S$$

follows the standard normal distribution, that is, $N(0, 1)$. Here

$$S^2 = p(1 - p)[1/n_1 + 1/n_2]$$

where

$$p = (k_1 + k_2)/(n_1 + n_2)$$

is the combined sample proportion and $p_i = k_i/n_i$ is the sample proportion. As a guide, the samples should be large ($n > 30$) for the assumption of normality to hold, and the proportions should not be less than 0.1 or greater than 0.9.

Example

Local contractors claim that, over a 10-year period, foreign contractors have been increasing their share of the local market for major projects. Two samples of projects, each of size 40, were taken 10 years apart and the results are given below:

Year	Sample size	Foreign	Local	p_i
2006	40 projects	20	20	20/40 = 0.500
2016	40 projects	25	15	25/40 = 0.625

It appears that foreign contractors are securing more projects, increasing their share from 50 percent to 62.5 percent. The issue is whether the difference is statistically significant. The combined sample proportion for contracts won by foreign contractors is

$$p = (20 + 25)/(40 + 40) = 0.563.$$

Thus,

$$S^2 = p(1 - p)[1/n_1 + 1/n_2] = 0.563(1 - 0.563)[1/40 + 1/40] = 0.0123.$$

Hence $S = \sqrt{0.0123} = 0.111$, and

$$Z = (0.625 - 0.5)/0.111 = 1.13.$$

From statistical tables, the 0.05 critical Z value (one-tailed) is 1.645. We do not reject H_0. There is no statistical evidence to show that foreign contractors have increased their share of major projects.

Repeat Measures Design

Paired t test

Recall from Chapter 7 that, in a repeat measures design, each experimental unit is measured twice. This differs from the parallel group design where there are two independent groups and we measure each experimental unit only once.

As before, we assume that the population is normally distributed. Under the null hypothesis of no difference in test scores, the test statistic is

$$t_{n-1} = m/(s/\sqrt{n}).$$

Here m is the mean difference in scores, s is the standard deviation of d, the difference in scores, and n is the sample size.

Example

In Table 13.1, 25 students took a mathematics test before they were exposed to a new teaching method, and then took a second test. The test scores are based on 0–100 marks. We then compute the differences in scores (d), its mean (m), and standard deviation (s).

Under the null hypothesis of no difference in scores, the test statistic is

$$t_{24} = m/(s/\sqrt{n}) = 4/(5/\sqrt{25}) = 4.0.$$

From Appendix 2, the 0.05 critical value (2-tailed) for t_{24} is 2.064. We reject H_0 and conclude that there is a statistical difference in scores.

Table 13.1 Data for teaching experiment.

Individual	Test scores Before	Test scores After	Difference d
1	90	60	−30
2	70	70	0
3	60	80	20
····	···	···	···
25 = n	80	90	10
			$m = 4$; $s = 5$

Randomized Block Design

The repeat measures design is, in fact, a form of blocking where each subject acts as his or her own control.

More generally, recall from Chapter 7 that we may wish to block out certain effects such as gender. That is, from a random sample of 100 students comprising 60 boys and 40 girls, we assigned them to experimental (E) and control (C) groups as shown in Table 13.2, which is a replica of Table 7.6.

The experimental groups were taught using a new method while the control groups were taught using the usual method. The data are shown in Table 13.3. The first group of boys (Students 1 to 30) were exposed to the new teaching method ($T = 1$). Boys are coded 1 for Gender, while girls are coded 0. The second group of boys (Students 31 to 60) were not exposed to the new teaching method ($T = 0$). For the girls, Students 61 to 80 were exposed to the new teaching method while Students 81 to 100 were taught the usual way.

In the past, the standard way to analyze the data was to use the traditional analysis of variance (for example, Milton and Arnold, 1995), or ANOVA. A simpler and more flexible way is to use the linear model (for example, Rawlings *et al.*, 1998) and define

$$Y_i = \alpha + \beta G_i + \lambda T_i + \phi G_i T_i + \varepsilon_i. \tag{13.1}$$

Here Y_i is the math score for the ith student, T is a dummy variable representing the new teaching method, with $T = 1$ if it is administered, and 0 otherwise. The variable G is the gender dummy variable with $G = 1$ if the student is a boy and 0 if the student is a girl. The interacting term GT is the product of G and T, which may also be written as $G(T)$ or $G \times T$ to indicate more clearly that it is a product of two variables. We shall use GT for simplicity. Finally, α is a constant, β, λ, and ϕ are parameters (coefficients), and ε is the error term.

Table 13.2   Illustration of blocking.

	60	30	E
100		30	C
	40	20	E
		20	C

Table 13.3　Data for randomized block design.

Student	Score (Y_i)	Teaching method (T_i)	Gender (G_i)
1	90	1	1
2	82	1	1
...	...	...	...
30	56	1	1
31	80	0	1
32	70	0	1
...	...	...	...
60	54	0	1
61	67	1	0
62	93	1	0
...	...	...	...
80	56	1	0
81	78	0	0
82	80	0	0
...	...	...	...
100	58	0	0

The interactive term captures the possibility that the new teaching method may affect boys and girls differently. For boys ($G = 1$), the above equation becomes

$$Y_i = \alpha + \beta + \lambda T_i + \phi T_i + \varepsilon_i = (\alpha + \beta) + (\lambda + \phi)T_i + \varepsilon_i.$$

For girls ($G = 0$), the estimating equation is

$$Y_i = \alpha + \lambda T_i + \varepsilon_i.$$

The intercepts as well as the slope coefficients differ between the two models. If the interacting term is not used, only the intercepts differ.

There is no need to estimate the equations separately for boys and girls. Hence, we just need to estimate Equation (13.1) using regression analysis. We will discuss regression analysis in the next chapter, but more so in the context of regression design rather than experimental design.

Interesting, the data analyses are similar for both experimental and regression designs. This is not unexpected; recall that regression designs are a form of experimental designs that use statistical control in place of experimental control. The advances in the analysis of linear models have made it possible to apply regression analysis flexibly to many types of experimental data. Finally, observe that, for modern ANOVA using experimental data, the right hand side consists of only ones and zeros (see Table 13.3).

References

Milton, J. and Arnold, J. (1995) *Introduction to probability and statistics*. New York: McGraw-Hill.

Rawlings, J., Pantula, S. and Dickey, D. (1998) *Applied regression analysis*. New York: Springer.

CHAPTER 14

Quantitative Data Analysis III: Regression Data (Part I)

Linear Regression

The simple regression model postulates that a *dependent variable Y* is a *linear* function of an *independent variable X*, that is,

$$Y_i = \alpha + \beta X_i + \varepsilon_i, \quad i = 1, 2, ..., n. \tag{14.1}$$

Here α is the constant term or intercept, β is the regression coefficient or parameter, ε_i is the *i*th error or disturbance term, and n is the number of data points. For example, we may postulate that monthly household expenditure (Y) depends on disposal income (X). For each family, we observe Y_i and X_i, giving a total of n data points in a (X, Y) scatter diagram.

We seldom use the simple regression model in actual research because it contains only one independent variable. In many cases, there are other variables such as household size, stage of life cycle, tastes, and so on. In such cases, a multiple regression model is used. We use the simple regression model mostly for expositional purposes. We can easily extend it into the multiple regression case.

For brevity, the subscripts are sometimes dropped so that

$$Y = \alpha + \beta X + \varepsilon.$$

Equation (14.1) is the *population* regression model. In practice, we take a sample and estimate

$$Y = a + bX + e. \tag{14.2}$$

Here a and b are the estimated intercept and coefficient respectively, and e is the *residual*, the sample estimate of ε. If we shift e to the left,

$$Y_i - e_i = Y_i^* = a + bX_i. \tag{14.3}$$

139

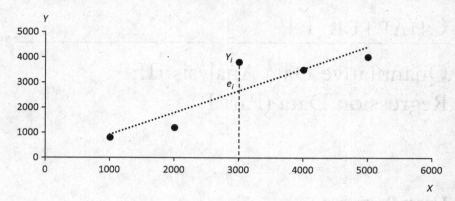

Fig. 14.1.   Sample regression line.

Here Y_i^* is a point *on* the estimated regression line. To recap, Y_i is the observed *i*th data point (i.e., the dots in Fig. 14.1), Y_i^* (not shown) is the corresponding point *on* the estimated regression line, and e_i is the difference $(Y_i - Y_i^*)$.

Model Assumptions

In estimating the regression line, we have implicitly used various assumptions. These assumptions are discussed below. Departures from these assumptions are common, and they require careful analysis.

1 *Normality*

We assume that Y, the dependent variable, is normally distributed. This assumption is required to facilitate the derivation and use of many statistical tests based on the normal distribution. This assumption is reasonable because many variables are normally distributed. However, there are also variables that are not normally distributed. For example, if we let the dependent variable S_i be whether a person is a smoker, we can regress it on independent variables such as age, income, lifestyle, education level, and stress level. Then S_i takes the value of 1 (smoker) or 0 (non-smoker). This dichotomous dependent variable is not normally distributed. Other examples of dichotomous dependent variables include like/dislike, yes/no, own/

rent, and agree/disagree. These variables are coded 1 or 0, and such cases are dealt with using the logit or probit regression models (see Chapter 15).

2 Linearity

The model is assumed to be linear, and

$$E(Y_i|X_i) = \alpha + \beta X_i.$$

This equation is the population regression line. Here E(.) is the expectation or mean, and $E(Y_i|X_i)$ is the expected or mean value of Y_i given X_i. Equivalently, it is the expected value of Y_i conditional on X_i.

The assumption of linearity implies that

$$E(\varepsilon_i) = 0.$$

That is, the errors are, on average, zero.

The linear assumption breaks down if the relation between the variables is nonlinear. In many cases, it is possible to transform a nonlinear function into a linear one. However, there are cases where it is not possible. We will discuss this issue in the next chapter.

3 Independence

If Y_i and Y_j are independent, the expenditure of different households do not affect each other. This assumption may break down if people spend to keep up with their neighbors, such as the purchase of luxury cars and other items.

From Equation (14.1), the independence assumption implies that the error terms are also uncorrelated, that is,

$$Cov(\varepsilon_i, \varepsilon_j) = 0$$

where Cov(·) is the covariance.

4 Distribution of errors

We further assume that

$$Var(Y_i|X_i) = \sigma^2.$$

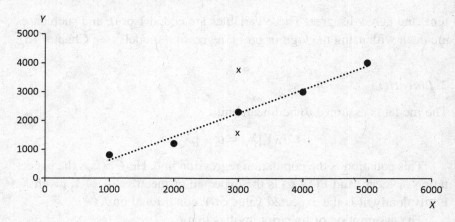

Fig. 14.2   Distribution of expenditure among households with income of $3,000.

This means that if we take a sample of households with the *same* income (for example, $3,000 in Fig. 14.2) and plot their monthly consumption expenditure (shown as crosses), then:

- the distribution of Y_i, given X_i, is normal (Assumption 1);
- the expected value of Y_i is $\alpha + \beta X_i$ (Assumption 2); and
- the variance of Y_i, given X_i, is σ^2.

Hence, ε_i is distributed as $NID(0, \sigma^2)$ where NID stands for normally and independently distributed. The assumption of constant variance or standard deviation (σ) will break down if households with higher incomes tend to vary more in their spending.

The error term is unobservable (i.e., unknown) and represents:

- the influences of omitted variables, presumably because of our ignorance;
- measurement errors; and
- random variations, such as in human behavior.

For these reasons, households with the same income level will vary in their spending on goods and services.

5 *X is fixed in repeated sampling*

The fifth assumption is that X is fixed (non-random), that is, it does not vary from sample to sample. If X is fixed, then

$$E(X_i \varepsilon_i) = X_i E(\varepsilon_i) = 0,$$

that is, they are uncorrelated. This simplifies the analysis; in particular, it is easy to prove that the ordinary least square (OLS) estimator is unbiased (see Chapter 15).

6 *n > k and linear independence*

This assumption is technical. If $n = 1$, there is only one data point, which makes it impossible to estimate the simple regression line. Similarly, if $n = 2$, there are only two data points and the regression line will have no error. Hence, a minimum of three data points is required to estimate the simple regression line using OLS.

In the case of multiple regression, such as

$$Y = \alpha + \beta X + \lambda Z + \varepsilon,$$

the variables X and Z must not be collinear, that is, they must be linearly independent. If Z is collinear or perfectly correlated with X, then Z is a linear function of X, such as $Z = \theta X$ for some constant θ, and

$$Y = \alpha + \beta X + \lambda \theta X + \varepsilon = \alpha + (\beta + \lambda \theta)X + \varepsilon.$$

In other words, the inclusion of Z does not add any information because the multiple regression model has become a simple regression equation. A more technical reason is that, in the presence of perfect collinearity, the model $Y = \alpha + \beta X + \lambda Z + \varepsilon$ cannot be estimated using OLS, and this is explained in the next chapter.

The more common problem is when Z is *highly correlated* with X, that is, they are *multicollinear* rather than perfectly collinear. For example, if Y is house price, X is land area, and Z is the number of rooms, then X and Z are highly correlated. A large house is likely to have more rooms. In this case, the inclusion of Z does not add much information, and either X or Z is dropped from the model.

Least Squares Estimation

The regression model in Equation (14.1) is usually estimated OLS. Other more complicated approaches, such as the maximum likelihood method, Bayesian method, robust regression or method of moments (see Birkes and Dodge, 1993) are used if some of the model assumptions are violated.

As an example, consider the model

$$Y_i = a + bD_i + cL_i + e_i; \quad i = 1, ..., n.$$

Here Y is house price, D is distance to the Central Business District (CBD), and L is land area. The data table for a small sample of 8 houses (for illustrative purposes) is shown in Table 14.1. In practice, we can add other variables such as amenities, tenure, and age of house as additional columns.

Nowadays, we use standard software such as Microsoft EXCEL, Minitab, Statistical Analysis System (SAS), or Statistical Package for the Social Sciences (SPSS). The estimated regression model is

$$Y^* = 251.3 - 27.59D + 2.99L \qquad R^2 = 0.98.$$

$$(\pm 4.72) \quad (\pm 0.56)$$

Before interpreting the model, we need to conduct some diagnostic tests. There are many such tests to check if the assumptions of the model hold. Some of these tests are discussed in the next chapter.

Table 14.1　Sample of eight houses and their characteristics.

[i]	House price Y ($'000)	Distance to CBD D (km)	Land area L (m²)
1	500	10	180
2	600	8	190
3	700	3	190
4	800	3	210
5	650	7	195
6	900	2	240
7	750	4	190
8	950	2	250

Coefficient of Determination

How well does the sample regression line fit the data? To answer this, we use the *coefficient of determination*

$$R^2 = 1 - RSS/TSS.$$

Here RSS is the sum of squares of the residuals, that is,

$$RSS = \Sigma e_i^2.$$

TSS is the total sum of squares, that is,

$$TSS = \Sigma(Y_i - m)^2$$

where m is the mean of the sample of Y values. If the sample regression line fits the data well, the residuals will be small. Thus, RSS/TSS will be close to zero and R^2 will be close to one. Hence, R^2 provides a measure of goodness of fit. A value of R^2 close to 1 indicates a good fit.

We cannot compare R^2 between two models if:

- the dependent variables are different, such as Y and log Y; or
- sample sizes are not identical.

In both cases, the ratio RSS/TSS will not be the same. Further, it is possible to improve the R^2 by discarding points that do not fit the line well. Thus, R^2 is a property of the sample, not of the population. For this reason, researchers should not take R^2 too seriously. Finally, because we can improve R^2 by adding more independent variables, a better measure of fit is the adjusted R^2, which is given by

$$R_a^2 = 1 - [(1 - R^2)(n - 1)/(n - k)].$$

For our model, $R^2 = 0.98$, $n = 8$, and $k = 3$. Hence $R_a^2 = 1 - [(0.02)(7)/5] = 0.97$. This suggests a good fit.

Tests of Significance

Recall that our estimated model is

$$Y^* = 251.3 - 27.59D + 2.99L \qquad\qquad R^2 = 0.98.$$
$$(\pm 4.72) \quad (\pm 0.56)$$

The regression model, in terms of population parameters, is

$$Y_i = \alpha + \beta D_i + \lambda L_i + \varepsilon_i.$$

We may use the estimated coefficients b and c, which are -27.59 and 2.99 respectively, to test hypotheses about β and λ. The intercept α is often of little interest, so we first test

$$H_0: \beta = 0 \text{ against } H_1: \beta \neq 0.$$

We test for $\beta = 0$ because if H_0 is true, then D does not affect Y.

To use b to test hypotheses about β, we need to find the distribution of b and develop a test statistic. Because Y is normally distributed, b is also normally distributed (Draper and Smith, 1998; see also Chapter 15). However, the standard deviation of b is unknown. What we have is the sample estimate (s_b), which is ± 4.72. Hence, we can use the Student t distribution and compute, under H_0,

$$t_{n-k} = b/s_b = -27.59/4.72 = -5.85.$$

Since $k = 3$, there are $n - k = 8 - 3 = 5$ degrees of freedom. The 0.05 critical value for t_5 value is 2.571 (two-tailed) and we reject H_0. This means that distance to CBD (D) has a significant effect on house price, which is not surprising.

Similarly, to test

$$H_0: \lambda = 0 \text{ against } H_1: \lambda \neq 0,$$

the corresponding t value is

$$t_5 = (2.99 - 0)/0.56 = 5.34.$$

We reject H_0 and conclude that land area also affects house price.

How do we interpret the results? The constant term (251.3) represents the base value of the house not accounted for by location and land area. It is the estimated value of Y when D and L are both zero. It therefore measures the contribution to house price by other house characteristics omitted from the model.

Recall from Chapter 8 that b represents the effect of a unit change in D on the expected house price. Hence, b ($= -27.59$) represents the expected *decrease* in house price with each kilometer from the city center, holding other variables that affect house price constant. In other words, all else equal, each kilometer from the city center *lowers* house price by an average of $27,590.

Similarly, on average, each square meter of land *contributes* $2,990 to house price, holding other variables are constant.

Forecasting

Once we are satisfied that the model is a reasonable one, it may be used to forecast the price for a house located, say 5 km, from the CBD with a land area of 200 m^2:

$$Y_F^* = 251.3 - 27.59(5) + 2.99(200) = \$711,350.$$

The forecast is an average value. There will be price variations among houses with the above characteristics.

How good is the forecast? It is likely to be unreliable because we use only two independent variables. Finally, out of sample forecasts well beyond the city fringe are unreliable because we did not use data beyond the city in our regression analysis. There are other limitations of the hedonic price model (see Chapter 8).

References

Birkes, D. and Dodge, Y. (1993) *Alternative methods of regression*. New York: Wiley.

Draper, N. and Smith, H. (1998) *Applied regression analysis*. New York: Wiley.

CHAPTER 15

Quantitative Data Analysis III: Regression Data (Part II)

The Linear Model in Matrix Form

This chapter extends the regression model in Chapter 14 to consider the various departures from ordinary least squares (OLS) assumptions. We will use simple regression to illustrate the principles because extending them to multiple linear regression presents little difficulty.

It is impossible to cover all topics; hence, we will only cover a selection of common issues. The exposition is introductory to create awareness of the issues. Readers can explore the statistical details by consulting specialized textbooks such as Long (1997), Seber and Lee (2003), and Monahan (2008).

We start with the linear regression model in matrix form

$$\mathbf{y} = \mathbf{X}\boldsymbol{\beta} + \boldsymbol{\varepsilon} \qquad (15.1)$$

where $\mathbf{y}$ is an $n \times 1$ vector of observations, $\mathbf{X}$ is the $n \times k$ data or design matrix, $\boldsymbol{\beta}$ is the $k \times 1$ vector of fixed parameters, and $\boldsymbol{\varepsilon}$ is the $n \times 1$ vector of error terms. We assume that the model is correctly specified, that is, $\mathbf{y}$ and $\mathbf{X}$ are linearly related.

We further assume that $\boldsymbol{\varepsilon} \sim N(\mathbf{0}, \sigma^2 \mathbf{I})$ where $\mathbf{I}$ is the $n \times n$ identity matrix. This is the matrix version of the error assumption that $\varepsilon_i \sim NID$ $(0, \sigma^2)$ where NID stands for normally and independently distributed. This means that the average error is zero, the error variance is constant (σ^2), and the errors are independent. The last assumption implies observations Y_i and Y_j for all i and j are also independent.

The independent variables are exogenous or fixed in repeated sampling, that is, they do not vary from sample to sample. In other words,

$$E(X_i \varepsilon_i) = X_i E(\varepsilon_i) = 0.$$

Further, we assume that there are more observations than parameters so that $n > k$. Otherwise, we do not have sufficient data points to fit the regression line. Finally, we require that the columns of $\mathbf{X}$ are linearly independent, that is, if $\mathbf{x}_i$ and $\mathbf{x}_j$ are any two column vectors of $\mathbf{X}$, their linear combination

$$\lambda \mathbf{x}_i + \theta \mathbf{x}_j = 0$$

implies $\lambda = \theta = 0$. If these scalars are not zero, we can write

$$\mathbf{x}_i = -(\theta/\lambda)\mathbf{x}_j.$$

That is, the column vectors $\mathbf{x}_i$ and $\mathbf{x}_j$ are parallel (collinear), or one vector is a scalar multiple of another.

For our sample, we can write

$$\mathbf{y} = \mathbf{Xb} + \mathbf{e} \tag{15.2}$$

where $\mathbf{b}$ is the ordinary least squares (OLS) estimator of β and $\mathbf{e}$ is the residual vector. To derive the OLS solution, we pre-multiply both sides by $\mathbf{X}^T$ so that

$$\mathbf{X}^T\mathbf{y} = \mathbf{X}^T\mathbf{Xb} + \mathbf{X}^T\mathbf{e}.$$

Here $\mathbf{X}^T$ is the transpose matrix of $\mathbf{X}$. It is a matrix obtained by rewriting each column of $\mathbf{X}$ as its rows. For example, we write the first column of $\mathbf{X}$ as the first row of $\mathbf{X}^T$, the second column of $\mathbf{X}$ as the second row of $\mathbf{X}^T$, and so on.

The OLS solution imposes the condition that the residual vector $\mathbf{e}$ is orthogonal to the space spanned by the columns of $\mathbf{X}$. This implies that

$$\mathbf{X}^T\mathbf{e} = 0$$

so that

$$\mathbf{X}^T\mathbf{y} = \mathbf{X}^T\mathbf{Xb}. \tag{15.3}$$

This gives a set of *normal equations*. Hence, the OLS estimator is

$$\mathbf{b} = (\mathbf{X}^T\mathbf{X})^{-1}\mathbf{X}^T\mathbf{y}. \tag{15.4}$$

The normal equations have a unique solution only if the matrix inverse exists. This means that the columns of $\mathbf{X}$ must be linearly independent. Equivalently, the determinant of $\mathbf{X}^T\mathbf{X}$ must not be zero. It is still possible to solve the normal equations using the generalized inverse if $\mathbf{X}^T\mathbf{X}$ is not invertible but the solution will not be unique. This approach is beyond the scope of this book, and readers can consult Rao (1965) and Searle (1971).

Properties of OLS Estimator

We now consider some properties of the OLS estimator. Before we proceed, it is important to know the difference between an estimator and an estimate. An estimator is a formula based on sample values. For example,

$$m = \Sigma x_i/n$$

is a formula for calculating the sample mean based on n sample values $x_1,\ldots,x_n$. Hence, m is an *estimator* of the population mean μ. The computed value, such as $m = 4.5$, is an *estimate*. There are many estimators of μ. For example,

$$m^* = (x_{min} + x_{max})/2$$

is also an estimator of μ. It takes the average of the minimum and maximum values of x to estimate the center of data. Why, then, do we prefer to use m rather than m^* as an estimator of μ?

The short answer is that m has better properties than m^*. Generally, we compare three properties, namely:

- unbiasedness;
- efficiency; and
- sufficiency.

An estimator is *unbiased* if its expectation equals its population parameter. In our example, it is μ. For estimator m,

$$E[m] = E[\Sigma x_i/n]$$

$$= \Sigma E[x_i]/n = n\mu/n = \mu.$$

The second line uses the result $E[x_i] = \mu$, and the expectation of a sum is the sum of individual expectations.

Similarly,

$$E[m^*] = E[(x_{min} + x_{max})/2]$$
$$= 2\mu/\mu = \mu.$$

Hence, both estimators are unbiased.

An estimator is *efficient* if it has lower variance. In our example,

$$Var[m] = Var[\Sigma\, x_i/n]$$
$$= \Sigma\, Var(x_i)/n^2$$
$$= n\sigma^2/n^2 = \sigma^2/n.$$

The second line uses the result that the variance of a sum is the sum of individual variances provided the xs are independent.

Similarly,

$$Var[m^*] = Var[(x_{min} + x_{max})/2]$$
$$= 2\sigma^2/4 = \sigma^2/2.$$

Comparing the two variances, Var (m) is more efficient because Var$(m) <$ Var(m^*) for sample sizes $n > 3$.

An estimator is *sufficient* if it uses all the information in the sample. In our example, m uses all the sample values, whereas m^* uses only two sample values, namely, the lowest and highest values of x.

In summary, both estimators are unbiased. However, m is more efficient, and uses more sample information than m^*. Hence, we prefer m to m^* as an estimator of μ. In practice, we usually compare estimators in terms of bias and efficiency. It is more difficult to prove sufficiency, which is why this criterion is seldom used.

We now return to the OLS estimator **b** and consider its bias and efficiency. It is unbiased because

$$E(\mathbf{b}) = E[(\mathbf{X}^T\mathbf{X})^{-1}\mathbf{X}^T\mathbf{y}]$$
$$= E[(\mathbf{X}^T\mathbf{X})^{-1}\mathbf{X}^T(\mathbf{X}\mathbf{b} + \mathbf{e})] = \beta.$$

Observe that in taking expectations above, we assume **X** is fixed. Otherwise, we cannot show that **b** is unbiased.

Further,

$$\mathbf{b} - \boldsymbol{\beta} = (\mathbf{X}^T\mathbf{X})^{-1}\mathbf{X}^T\mathbf{y} - \mathbf{b}$$
$$= (\mathbf{X}'\mathbf{X})^{-1}\mathbf{X}^T(\mathbf{X}\boldsymbol{\beta} + \boldsymbol{\varepsilon}) - \boldsymbol{\beta} = (\mathbf{X}^T\mathbf{X})^{-1}\mathbf{X}^T\boldsymbol{\varepsilon}.$$

Thus,

$$\text{Var}(\mathbf{b}) = E[(\mathbf{b} - \boldsymbol{\beta})(\mathbf{b} - \boldsymbol{\beta})^T]$$
$$= E[(\mathbf{X}^T\mathbf{X})^{-1}\mathbf{X}^T\boldsymbol{\varepsilon}\,((\mathbf{X}^T\mathbf{X})^{-1}\mathbf{X}^T\boldsymbol{\varepsilon})^T] = \sigma^2(\mathbf{X}^T\mathbf{X})^{-1}. \tag{15.5}$$

From Equation (15.4), we see that $\mathbf{b}$ is a linear function of $\mathbf{y}$. If the distribution of $\mathbf{y}$ is normal, then the distribution of $\mathbf{b}$ is also normal.

In terms of efficiency, $\mathbf{b}$ is the best linear unbiased estimator (BLUE). That is, among all linear unbiased estimators, it has the smallest variance, which is a desirable property. For a proof of this result, see Maddala (1986).

In summary, the OLS estimator $\mathbf{b}$ has a normal distribution with mean $\boldsymbol{\beta}$ and variance given by Equation (15.5). It has two desirable properties, namely, it is unbiased and, among all linear unbiased estimators, it is best in the sense that it has the smallest variance. There are *biased* estimators that have smaller variances than $\mathbf{b}$, but this is beyond the scope of this book. Interested readers may consult Gruber (1998).

Tests of Significance

We can use the distribution of $\mathbf{b}$ to test for the significance of each regression coefficient. However, in Equation (15.5), σ^2 is unknown and we have to estimate it from the residuals, that is,

$$s^2 = \mathbf{e}^T\mathbf{e}/(n - k). \tag{15.6}$$

We use Equations (15.5) and (15.6) to test the significance of each element of $\boldsymbol{\beta}$ (for example, β_i). That is, under H_0: $\beta_i = 0$,

$$b_i/SD(b_i) \sim t_{n-k}$$

where $SD(.)$ denotes the standard deviation. This is the usual t test on the significance of each regression coefficient (see Chapter 14).

Transformation of Nonlinear Functions

We start with the assumption that the linear model in Equation (15.1) is correctly specified. Often, it is possible to transform a nonlinear function into a linear one. For example, we can linearize the cost function

$$C = \alpha + \beta Q + \lambda Q^2$$

by letting $T = Q^2$ so that

$$C = \alpha + \beta Q + \lambda T.$$

Here C is the cost of production, and Q is output. We say that the function is nonlinear in the variable but linear in the parameters. A function is nonlinear in the parameters if it contains powers and products of parameters. For example, the function

$$Y = \alpha + \beta \theta Q + \lambda^2 T$$

is nonlinear in the parameters.

Next, consider a multiplicative model

$$Z = A S^\lambda E^\beta$$

where Z is wage rate, A is a constant, S is years of schooling, and E is years of experience. We can transform it into linear form by taking natural logs on both sides so that

$$\log(Z) = \log(A) + \lambda\log(S) + b\log(E).$$

By adding an error term (ε), we can use OLS to regress $\log(Z)$ against $\log(S)$ and $\log(E)$ by letting $z = \log(Z)$, $s = \log(S)$, and $e = \log(E)$, that is,

$$z = \alpha + \lambda s + \beta e + \varepsilon.$$

The intercept is $\log(A)$, which we can designate as α. To obtain A, use the antilog function, that is, $A = \text{antilog}(\alpha)$. We must remember to convert the raw data Z, S, and E into logs before estimating the regression model.

A trickier problem is to transform the logit function

$$Y_t = [1 + e^{-(\alpha + \beta t)}]^{-1}$$

into linear form. The logit function is an "S curve" with values of Y on the vertical axis and time (t) on the horizontal axis (Fig. 15.1).

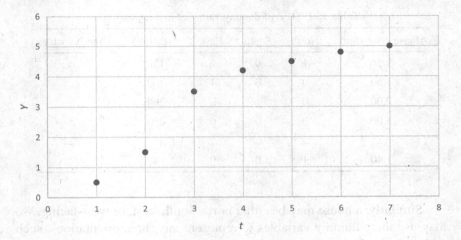

Fig. 15.1   A logistic curve.

For example, it may represent how project cost (Y) varies with the duration of the project.

To carry out the linear transformation, we first write it as

$$1/Y_t = 1 + e^{-(\alpha + \beta t)}$$

so that

$$\log[1/Y_t - 1] = -(\alpha + \beta t) = \lambda_1 + \lambda_2 t.$$

By letting $Z = \log[1/Y_t - 1]$, we now have a linear model

$$Z = \lambda_1 + \lambda_2 t.$$

In summary, I have shown how to transform some common functions into linear models so that we can estimate these equations using OLS.

Dummy Variables

We use dummy variables in a regression to capture the effects of nominal variables on the dependent variable. For example, in the house price model, we may introduce a dummy variable GARAGE to indicate whether a house has a garage. Here GARAGE is coded 1 if a house has a garage, and 0 otherwise.

Table 15.1   Use of dummy variables in a regression.

House price ($m)	North	South	East	Area (m²)	Other variables
1.56	1	0	0	150	...
1.60	0	0	1	160	...
2.00	0	1	0	200	...
3.00	0	0	0	250	...
...	...	...	...	...	...
1.40	1	0	0	140	...

Similarly, a house may be either north, south, east, or west-facing. We may use three dummy variables to represent any three orientations, such as North, South, and East (Table 15.1). The first house faces north, the second house faces east, the third house faces west, and so on.

We write the model as

$$P = \alpha + \beta\text{North} + \lambda\text{South} + \theta\text{East} + \Upsilon\text{Area} + \ldots + \varepsilon.$$

In general, we use $c - 1$ dummy variables to represent c categories. There are two categories in GARAGE (garage, no garage), so we use one dummy variable, GARAGE. There are four categories in orientation (North, South, East, and West), so we can use any three dummy variables.

Why do we use $c - 1$ dummy variables to represent c categories? The reason is that the columns of the data matrix will be collinear if we use c dummy variables. To understand this, consider a simple model where a student's mathematics score (S) depends on gender and monthly household income (C). If we use two variables Male and Female to represent gender (Table 15.2), the model is

$$S = \alpha + \beta\text{Male} + \lambda\text{Female} + \theta C + \varepsilon.$$

Observe that, for each row, the sum of the values for the Male and Female variables equals 1. If we let x_1 be the first column vector of the data matrix X (see Equation (15.1)), x_2 be the second column vector consisting of binary data for the Male variable, and x_3 be the third column vector consisting of binary data for the Female variable, then

$$x_1 = x_2 + x_3.$$

Table 15.2   Illustration of collinearity.

S	Male	Female	C($)
90	1	0	5,000
80	0	1	4,000
70	0	1	3,500
...	...	...	...
85	0	1	5,200

The first three column vectors of X are collinear, and the normal equations do not have a unique solution.

The effect of a dummy variable is to shift the regression line or, equivalently, change the intercept. For example, if

$$W = \alpha + \beta X + \lambda G$$

where W is wage rate, E is work experience, and G is gender ($G = 0$ for female and 1 for male), then the equations for males and females are

$$W = \alpha + \beta X + \lambda = (\alpha + \lambda) + \beta X \qquad \text{(males); and}$$
$$W = \alpha + \beta X \qquad \qquad \text{(females)}.$$

Observe that the two equations differ only in the intercept term. If $\lambda > 0$, then, all else equal, males are paid more than females. The situation is reversed if $\lambda < 0$. In other words, λ is an indicator of wage differences due to gender. Hence, a test of whether $\lambda = 0$ is critical for this wage model.

It is not necessary to regress the two equations separately. We just need to estimate the original regression model $W = \alpha + \beta X + \lambda G$ directly.

Interacting Variables

Let Y be a dichotomous variable where $Y = 1$ if a person has lung cancer, and 0 otherwise. Note Y is a dependent variable, and hence it is not a dummy independent variable. Let

$$Y = \alpha + \beta X + \lambda G + \theta XG + \varepsilon$$

where X is the number of cigarettes smoked per day, G is a gender dummy variable ($G = 0$ for female and 1 for male) and XG (that is,

X multiplied by G) is an *interacting term*. It is the product of two independent variables.

If we interpret $E(Y)$ as the probability of having lung cancer, then this probability depends on X, G, and XG. To see the need for an interacting term, we differentiate $E(Y)$ partially with respect to X so that

$$\partial E(Y)/\partial X = \beta + \theta G.$$

Thus, we are postulating that the change in $E(Y)$ resulting from the number of cigarettes smoked affects men and women differently. It does not make a difference only if $\theta = 0$.

In summary, the effect of each independent variable on the dependent variable Y may not be additive. The independent variables may interact to produce a change in $E(Y)$ arising from a change in X_i that, in turn, depends on the level of another variable X_j.

Maximum Likelihood Estimation

There are nonlinear equations such as

$$Y = f(X, \alpha, \beta) = \alpha X^\beta + \varepsilon \tag{15.7}$$

where it is not possible to transform into a linear model by simple transformations. Such equations are *intrinsically nonlinear*.

There are two main methods of estimating intrinsically nonlinear models. We first discuss maximum likelihood estimation (MLE) here and then nonlinear least squares (NLS) in the next section.

If the error terms are normally distributed, the probability density function is

$$f(\varepsilon_i) = (2\pi\sigma^2)^{-1/2}\exp(-\varepsilon_i^2/2\sigma^2)$$

where $\exp(.)$ is the exponential function, that is, $\exp(x) = e^x$ where e is the base of natural logarithm. From Equation (15.7),

$$\varepsilon_i = Y_i - \alpha X_i^\beta$$

so that

$$d\varepsilon_i/dY_i = 1.$$

Hence, the probability density function of Y_i is

$$g(Y_i) = f(\varepsilon_i)|d\varepsilon_i/dY_i| = (2\pi\sigma^2)^{-1/2}\exp[-(Y_i - \alpha X_i^\beta)^2/2\sigma^2].$$

Here $|d\varepsilon_i/dY_i|$ is the *Jacobian* of the density transformation. It scales the two density functions so that they are compatible. Effectively, we are deriving the density of our observations ($g(Y_i)$) by assuming that the errors are normally distributed.

For a sample of n observations $Y_1,\ldots, Y_n$, the *likelihood function* is the product of the individual density functions so that

$$L = g(Y_1)\ldots g(Y_n).$$

This function is usually difficult to optimize because each $g(Y_i)$, as shown above, contains a complicated exponential function. It is often easier to maximize the log likelihood function

$$Q = \log L = -n/2\log (2\pi\sigma^2) -(1/2\sigma^2)\Sigma(Y_i - \alpha X_i^\beta)^2.$$

To maximize this function, we partially differentiate Q with respect to each parameter and set it to zero:

$$\partial Q/\partial\alpha = -(1/2\sigma^2)\Sigma\, 2(Y_i - \alpha X_i^\beta)(-X_i^\beta) = 0;$$
$$\partial Q/\partial\beta = -(1/2\sigma^2)\Sigma\, 2(Y_i - \alpha X_i^\beta)(-\alpha X_i^\beta\log X_i) = 0;\ \text{and}$$
$$\partial Q/\partial\sigma^2 = -(n/2\sigma^2) + (1/2\sigma^4)\Sigma\, (Y_i - \alpha X_i^\beta)^2 = 0.$$

We solve these equations for the estimated parameters. Clearly, it is not easy to solve this set of nonlinear equations for α, β, and σ. This is why the second method, nonlinear least squares (NLS), is more commonly used.

In summary, the method of maximum likelihood seeks to estimate parameters that maximize the probability or likelihood of obtaining the sample. In practice, it is analytically easier to maximize the log likelihood function. Nonetheless, it is still difficult to solve the resulting equations for the parameters.

Nonlinear Least Squares

The NLS method linearizes the nonlinear model using Taylor series approximation about an initial set of parameters, called the starting values, and then applies OLS to the linear model. We then update the estimated parameters by applying OLS iteratively until the solution converges based on some convergence criteria.

For Equation (15.7), the Taylor series approximation about initial estimates α_0 and β_0 is

$$f(X, \alpha, \beta) = f(X, \alpha_0, \beta_0) + (\partial f/\partial \alpha)\Delta\alpha + (\partial f/\partial \beta)\Delta\beta.$$

We evaluate the partial derivatives at α_0 and β_0. If we shift the first term on the right hand side to the left, we obtain

$$\Delta f = (\partial f/\partial \alpha)\Delta\alpha + (\partial f/\partial \beta)\Delta\beta. \tag{15.8}$$

Here,

$$\Delta f = f(X, \alpha, \beta) - f(X, \alpha_0, \beta_0),$$
$$\Delta\alpha = \alpha - \alpha_0, \text{ and}$$
$$\Delta\beta = \beta - \beta_0.$$

Equation (15.8) is a linear model. In the first iteration, we regress Δf against $\partial f/\partial \alpha$ and $\partial f/\partial \beta$ to obtain estimates of $\Delta\alpha$ and $\Delta\beta$. Note that there is no intercept term in the equation.

For the second iteration, the updated estimates are

$$\alpha = \alpha_0 + \Delta\alpha; \text{ and}$$
$$\beta = \beta_0 + \Delta\beta.$$

For example, if we start with $\alpha_0 = 1$ and $\beta_0 = 2$ and obtain OLS estimates $\Delta\alpha = 0.1$ and $\Delta\beta = 0.2$, then the updated estimates are

$$\alpha = \alpha_0 + \Delta\alpha = 1 + 0.1 = 1.1; \text{ and}$$
$$\beta = \beta_0 + \Delta\beta = 2 + 0.2 = 2.2.$$

We use these updated estimates for the second iteration. We repeat the process until the difference between successive estimates is negligible. However, the process may not converge to a solution if:

- the initial guess value is not good; or
- the surface is rather "flat" near the solution, making it difficult to find an optimal least squares point (Seber and Wild, 2003).

Example

Recall from Equation (15.7) that our nonlinear model is

$$Y = f(X, \alpha, \beta) = \alpha X^{\beta} + \varepsilon$$

For the data in the first two columns of Table 15.3, we can see from the plot in Fig. 15.2 that β is positive, so our initial guesses are

$$\alpha_0 = 1; \text{ and}$$
$$\beta_0 = 2.8.$$

Hence, using Equation (15.7) and ignoring the error term for the moment, we use

$$Y_0 = X^{2.8}$$

to compute the values of Y_0 in the third column.

Table 15.3   Computation of values for regression variables.

X	Y	Y_0	Δf	$\partial f / \partial \alpha$	$\partial f / \partial \beta$
0	0	0	0	0	0
1	2	1	1	1	2.8
2	7	6.96	0.04	6.96	9.75
2.5	14	13	1	9.75	14.6

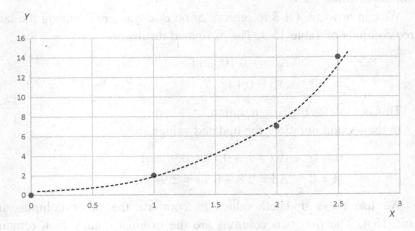

Fig. 15.2   Data for nonlinear least squares.

Next, we use the relation

$$\Delta f = Y - Y_0$$

to compute the values in the fourth column. Note that Δf is the difference between the observed value (Y) and the estimated value (Y_0) computed from our starting values. The relatively small differences indicate that the initial guess is good.

The next step is to find $\partial f/\partial\alpha$ and $\partial f/\partial\beta$ or, equivalently, $\partial Y/\partial\alpha$ and $\partial Y/\partial\beta$. Since

$$Y = \alpha X^\beta,$$

the partial derivatives are

$$\partial f/\partial\alpha = X^\beta; \text{ and}$$
$$\partial f/\partial\beta = \alpha\beta X^{\beta-1}.$$

For the first iteration, these partial derivatives are evaluated at the starting values $\alpha_0 = 1$ and $\beta_0 = 2.8$, giving

$$\partial f/\partial\alpha = X^{2.8}; \text{ and}$$
$$\partial f/\partial\beta = 2.8X^{1.8}.$$

The evaluated partial derivatives are shown in the fifth and sixth columns in Table 15.3.

We can now use OLS to regress Δf on $\partial f/\partial\alpha$ and $\partial f/\partial\beta$ using the last three columns of Table 15.3. The estimated results are

$$\Delta\alpha = -0.10; \text{ and}$$
$$\Delta\beta = 0.14.$$

This completes the first iteration.

In the second iteration, the updated values are:

$$\alpha = \alpha_0 + \Delta\alpha = 1 - 0.10 = 0.9; \text{ and}$$
$$\beta = \beta_0 + \Delta\beta = 2.8 + 0.14 = 2.94.$$

We use these updated values to compute the other columns in Table 15.4. The first two columns are the original data, which remain unchanged.

Table 15.4 Computation of new values for regression variables.

X	Y	Y_0	Δf	$\partial f/\partial\alpha$	$\partial f/\partial\beta$
0	0	0	0	0	0
1	2	0.9	1.1	1	2.65
2	7	6.91	0.09	7.67	10.17
2.5	14	13.31	0.69	14.79	15.68

For the third column, Y_0 is computed using

$$Y_0 = 0.9X^{2.94}.$$

Next, we compute $\partial f/\partial\alpha$ and $\partial f/\partial\beta$ using

$$\partial f/\partial\alpha = X^\beta = X^{2.94}; \text{ and}$$
$$\partial f/\partial\beta = \alpha\beta X^{\beta-1} = 0.9(2.94)X^{1.94} = 2.65X^{1.94}.$$

Thus, in the second iteration, the regression results are

$$\Delta\alpha = -0.19; \text{ and}$$
$$\Delta\beta = 0.21.$$

These results are only slightly different from that of the first iteration, indicating we are close to convergence. The updated estimates for the parameters are

$$\alpha = \alpha_0 + \Delta\alpha = 0.9 - 0.19 = 0.71; \text{ and}$$
$$\beta = \beta_0 + \Delta\beta = 2.94 + 0.21 = 3.15.$$

If desired, we can carry out another iteration to obtain a closer approximation.

Non-Normal Errors

For the linear regression model

$$Y = \alpha + \beta X + \varepsilon,$$

it is assumed that

$$\varepsilon_i \sim N(0, \sigma^2)$$

so that diagnostic tests based on normal distribution may be constructed. Based on OLS assumptions, we fix the values of α, β, and X. Hence, if Y is not normally distributed, then we can conclude from the equation that ε is also not normally distributed. In such cases, it may be possible to transform Y so that its distribution, and hence that of the error term, is close to normal. Note that we are transforming the dependent variable Y, rather than the independent variable X, here. It is important to keep this point in mind because it can confuse many students.

Box and Cox (1964) proposed a family of transformations:

$$Y^{(\lambda)} = Y^{\lambda} \qquad \text{if } \lambda \neq 0;$$
$$= \log Y \quad \text{if } \lambda = 0.$$

The second line is required because the first transformation is trivial if $\lambda = 0$, that is, $Y^0 = 1$. Some texts use

$$Y^{(\lambda)} = (Y^{\lambda} - 1)/\lambda \quad \text{if } \lambda \neq 0$$

instead of simply $Y^{(\lambda)} = Y^{\lambda}$ if $\lambda \neq 0$ because $(Y^{\lambda} - 1)/\lambda$ tends towards $\log Y$ as λ tends towards 0. We will not use this more complicated version here.

The problem is to find λ. For example, if $\lambda = 2$, the model is

$$Y^2 = \alpha + \beta X + \varepsilon.$$

Similarly, if $\lambda = 0$, the model is

$$\log Y = \alpha + \beta X + \varepsilon.$$

We may use a search procedure and select the value of λ that minimizes the sum of squares of residuals (RSS). For instance, we may start with $\lambda = -1$, regress

$$Y^{-1} = \alpha + \beta X + \varepsilon,$$

and compute RSS ($\lambda = -1$). The next value may be $\lambda = 0$ so that we regress

$$\log Y = \alpha + \beta X + \varepsilon$$

and compute RSS ($\lambda = 0$), and so on. By plotting RSS against λ, we can find the value of λ that gives the minimum RSS.

In summary, if the distribution of Y is not normal, we may transform it so that it is approximately normal. The Box-Cox transformation provides one possible method. It requires us to conduct a grid search to find a suitable value of λ that will help us decide on the type of transformation required.

Outliers

The regression line is sensitive to remote data points (outliers) because the OLS estimator minimizes the sum of squares of residuals (RSS). This problem is particularly severe in small samples.

An outlier is a point remote from the main cluster of data points. In Fig. 15.3, the two points on the right, (3, 14) and (3, 2), are outliers. However, point (3, 14) does not significantly affect the regression line if we exclude point (3, 2) from the regression. In contrast, if we include point (3, 2), it will significantly alter the slope of the regression line. We say that point (3, 2) is *influential*.

Some outliers are gross errors, such as if we key in 18 instead of 81. If we can detect gross errors, it is possible to correct them. However, there may be outliers that are not gross errors. In such cases, it is inappropriate to remove them just because they do not fit the theory. Instead, we should use them to refute or refine the theory.

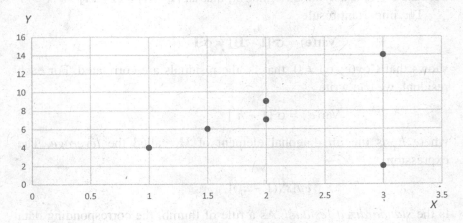

Fig. 15.3 Outlier and influential points.

If there are many data points and variables, it is difficult to identify outliers through graphical inspection. We have to examine the residuals for outliers in a formal statistical way. From Equation (15.2),

$$\mathbf{e} = \mathbf{y} - \mathbf{Xb} = \mathbf{y} - \mathbf{X}(\mathbf{X}^T\mathbf{X})^{-1}\mathbf{X}^T\mathbf{y} \quad \text{using the normal equations } \mathbf{b} = (\mathbf{X}^T\mathbf{X})^{-1}\mathbf{X}^T\mathbf{y}$$
$$= [\mathbf{I} - \mathbf{X}(\mathbf{X}^T\mathbf{X})^{-1}\mathbf{X}^T]\mathbf{y} \quad \text{where } \mathbf{I} \text{ is the identity matrix}$$
$$= [\mathbf{I} - \mathbf{H}]\mathbf{y} \quad \text{where } \mathbf{H} = \mathbf{X}(\mathbf{X}^T\mathbf{X})^{-1}\mathbf{X}^T \text{ is the } \textit{hat matrix.}$$

Hence,

$$\text{Var}(\mathbf{e}) = \text{Var}([\mathbf{I} - \mathbf{H}]\mathbf{y})$$
$$= [\mathbf{I} - \mathbf{H}]\text{Var}(\mathbf{y})[\mathbf{I} - \mathbf{H}]^T$$
$$= \sigma^2[\mathbf{I} - \mathbf{H}].$$

The second line above uses the relation

$$\text{Var}(\mathbf{By}) = \mathbf{B}\text{Var}(\mathbf{y})\mathbf{B}^T$$

where **B** is a matrix. Further,

$$\text{Var}(\mathbf{y}|\mathbf{X}) = \sigma^2\mathbf{I}.$$

Finally, $\mathbf{I} - \mathbf{H}$ is an *idempotent matrix*. A matrix is idempotent if $\mathbf{A}^2 = \mathbf{A}$, that is, its product is itself. Thus,

$$[\mathbf{I} - \mathbf{H}][\mathbf{I} - \mathbf{H}]^T = [\mathbf{I} - \mathbf{H}]$$

because $\mathbf{I} - \mathbf{H}$ is a symmetric matrix, that is, $[\mathbf{I} - \mathbf{H}] = [\mathbf{I} - \mathbf{H}]^T$.

The important result

$$\text{Var}(\mathbf{e}) = \sigma^2[\mathbf{I} - \mathbf{H}] \neq \sigma^2\mathbf{I}$$

shows that $\text{Cov}(e_i, e_j) \neq 0$, that is, the residuals are correlated. For each residual, we can write

$$\text{Var}(e_i) = \sigma^2[1 - h_{ii}]$$

where h_{ii} is the ith diagonal element of **H**, called the *leverage*. The expression

$$e_i/[s\sqrt{(1 - h_{ii})}]$$

is the *standardized residual*. As a rule of thumb, the corresponding data point may be an outlier if the above ratio exceeds 2.5. Note that the standardized residual does not follow the usual t distribution because e_i and s are not independent.

An alternative way of identifying outliers is to use the *Studentized residual*. Here, we replace s above with $s_{(i)}$, the standard deviation of the residuals from a regression *without* the ith outlier. It provides a better measure of σ than s because an outlier will inflate the value of s. We can observe this in Fig. 15.3 where the inclusion of the influential point (3, 2) in the regression will increase the value of RSS.

A simpler approach to detect outliers is to identify an observation as influential if

$$h_{ii} > 3k/n$$

where k is the number of parameters and n is the number of observations. The logic for this choice is that the average value of h_{ii} is k/n. The proof is simple:

$$\text{Trace } (\mathbf{H}) = \text{Trace } (\mathbf{X}(\mathbf{X}^\mathsf{T}\mathbf{X})^{-1}\mathbf{X}^\mathsf{T})$$
$$= \text{Trace } (\mathbf{X}^\mathsf{T}\mathbf{X}(\mathbf{X}^\mathsf{T}\mathbf{X})^{-1}) \quad \text{using Trace } (\mathbf{AB}) = \text{Trace } (\mathbf{BA})$$
$$= \text{Trace } (\mathbf{I}) = k.$$

The Trace (.) operator gives the sum of the diagonals of a matrix. If the sum of the diagonals of $\mathbf{H}$ is k, then its average value is k/n. For a more detailed discussion on outlier detection, see Belsley *et al.* (1980).

Testing Restrictions on Parameters

Consider the regression model

$$Y = \alpha + \beta X + \lambda T + \varepsilon.$$

Recall from Chapter 14 that we may test restrictions on a parameter such as β (for example, $\beta = 0$) using the t test. There are situations where we may want to test certain restrictions on more than one parameter together, such as restrictions on β and λ. For example, to test whether the overall regression model makes sense, we test

$$H_0: \beta = \lambda = 0,$$

that is, we have a *restricted* model

$$Y = \alpha + \varepsilon.$$

If the restriction is valid, both models should fit the data well. Hence, $RSS_R - RSS_U$ should be "small," where RSS_U is the residual sum of squares from the *unrestricted* model, and RSS_R is the residual sum of squares from the restricted model. To remove its dependence on the units of measurement, we can divide the difference in RSS by RSS_U together with their appropriate degrees of freedom. It is possible to show that (Johnston and DiNardo, 1997)

$$(RSS_R - RSS_U)/d \div RSS_U/(n-k) \sim F(d, n-k). \qquad (15.9)$$

As before, n is the number of observations, and k is the number of parameters in the unrestricted model. $F(.)$ is the F distribution with d and $n - k$ degrees of freedom respectively, where d is the number of restrictions. In this case, there are two restrictions, $\beta = 0$, and $\lambda = 0$. A large F value greater than the 0.05 critical value implies that the difference in RSS is "large," and we should reject H_0.

As a second example, we consider the case where we impose restrictions on the parameters to *test for structural change*. In Fig. 15.4, there appears to be a break in the data around $X = 5$, resulting in two possible regression lines. The first line uses the first five points, and the second line uses the last three points. Hence, we may write:

$$Y_1 = \alpha_1 + \beta_1 X + \varepsilon_1; \text{ and}$$
$$Y_2 = \alpha_2 + \beta_2 X + \varepsilon_2;$$

This is the unrestricted model. The structural break may occur because of a change in government policy that affects the relation between Y and X.

The restricted model is the regression line that uses all eight data points, that is,

$$Y = \alpha + \beta X + \varepsilon.$$

The null hypothesis is that there is no structural change, that is,

$$H_0: \alpha_1 = \alpha_2, \text{ and } \beta_1 = \beta_2.$$

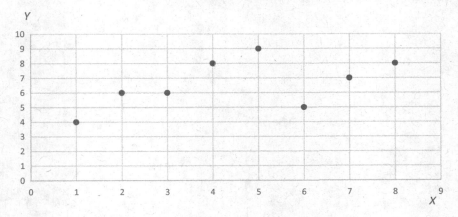

Fig. 15.4   Testing for structural change.

Thus, there are two restrictions. As before, we can use Equation (15.9) to test for structural change.

Heteroscedasticity

The assumption of homoscedasticity, that is, $\varepsilon_i \sim N(0, \sigma^2)$ or, in matrix form, $\varepsilon \sim N(0, \sigma^2 I)$, may not hold in practice. Consider the model

$$Y = \alpha + \beta X + \lambda L + \varepsilon$$

where Y is the rate of profit, X is firm size, and L is labor input. Because larger firms tend to have greater variability in the rate of profit compared to smaller ones, the errors or residuals may be bigger for larger firms (Fig. 15.5). We say that the errors are heteroscedastic.

There are other several ways to detect heteroscedasticity besides plotting Y against each independent variable, such as the Goldfeld-Quandt test or the White test. However, these tests are ad hoc (Maddala, 1986), and we will not consider them here.

To deal with heteroscedasticity, we may deflate (divide) Y by L so that

$$Y/L = \alpha + \beta X + \varepsilon.$$

The dependent variable is now Y/L, or profit per worker. This transformation "compresses" the errors to reduce the heteroscedasticity.

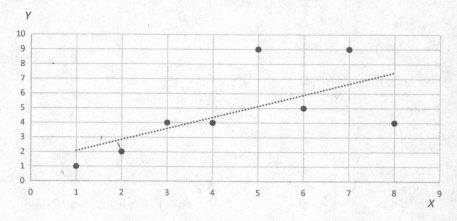

Fig. 15.5   Example of heteroscedasticity.

Alternatively, we may regress

$$\log(Y) = \alpha + \beta\log(X) + \lambda\log(L) + \varepsilon$$

and obtain a similar compressive effect on the errors.

A third approach is to model the heteroscedasticity explicitly. If the errors are heteroscedastic, then

$$\mathrm{Var}(\varepsilon) = \mathrm{E}[\varepsilon\varepsilon^T] = \sigma^2\Omega = \Sigma$$

where Ω (or Σ), called the covariance matrix, is no longer an identity matrix, that is, heteroscedasticity is present. The matrices $\sigma^2\Omega$ and Σ are equivalent, and which one to use is a matter of convenience. The diagonal elements of Ω (or Σ) are no longer ones but individual variances (i.e. σ_i^2). These variances are unequal. If the off-diagonal elements are zeros, then the covariance

$$\sigma_{ij} = 0$$

and Σ is a diagonal matrix. In this special case, we call the procedure *weighted least squares* (WLS). If

$$\sigma_{ij} \neq 0,$$

we call the procedure *generalized least squares* (GLS).

Consider a non-singular transformation matrix $\mathbf{T}$ such that

$$\mathbf{T}\mathbf{y} = \mathbf{T}\mathbf{X}\boldsymbol{\beta} + \mathbf{T}\boldsymbol{\varepsilon}.$$

That is,

$$\mathbf{z} = \mathbf{L}\boldsymbol{\beta} + \mathbf{u} \qquad (15.10)$$

where $\mathbf{z} = \mathbf{T}\mathbf{y}$, $\mathbf{L} = \mathbf{T}\mathbf{X}$, and $\mathbf{u} = \mathbf{T}\boldsymbol{\varepsilon}$. Then,

$$E[\mathbf{u}\mathbf{u}^\mathrm{T}] = E[\mathbf{T}\boldsymbol{\varepsilon}\boldsymbol{\varepsilon}^\mathrm{T}\mathbf{T}^\mathrm{T}] = \sigma^2\mathbf{T}\boldsymbol{\Omega}\mathbf{T}^\mathrm{T}.$$

We choose $\mathbf{T}$ such that $\mathbf{T}\boldsymbol{\Omega}\mathbf{T}^\mathrm{T} = \mathbf{I}$ so that OLS may now be applied to Equation (15.10). The normal equations are

$$\begin{aligned}
\mathbf{b}_{\mathrm{GLS}} &= (\mathbf{L}^\mathrm{T}\mathbf{L})^{-1}\mathbf{L}^\mathrm{T}\mathbf{z} \\
&= (\mathbf{X}^\mathrm{T}\boldsymbol{\Omega}^{-1}\mathbf{X})^{-1}\mathbf{X}^\mathrm{T}\boldsymbol{\Omega}^{-1}\mathbf{y} \\
&= (\mathbf{X}^\mathrm{T}\boldsymbol{\Sigma}^{-1}\mathbf{X})^{-1}\mathbf{X}^\mathrm{T}\boldsymbol{\Sigma}^{-1}\mathbf{y}. \qquad (15.11)
\end{aligned}$$

If $\boldsymbol{\Omega}$ is a diagonal matrix, then

$$\mathbf{W} = \boldsymbol{\Omega}^{-1}$$

is the *weight matrix* in the weighted least squares procedure. Finally,

$$\begin{aligned}
\mathrm{Var}(\mathbf{b}_{\mathrm{GLS}}) &= \sigma^2(\mathbf{L}^\mathrm{T}\mathbf{L})^{-1} \\
&= \sigma^2(\mathbf{X}^\mathrm{T}\boldsymbol{\Omega}^{-1}\mathbf{X})^{-1} \\
&= (\mathbf{X}^\mathrm{T}\boldsymbol{\Sigma}^{-1}\mathbf{X})^{-1}.
\end{aligned}$$

You should compare this result with Equation (15.5), where the errors are homoscedastic.

In summary, if the errors are heteroscedastic, we should use the GLS estimator in Equation (15.11). However, $\boldsymbol{\Omega}$ or $\boldsymbol{\Sigma}$ is often unknown, and we may have to estimate it from an OLS regression before using Equation (15.11).

Multicollinearity

Recall that it is not possible to solve the normal equations in Equation (15.4) if $\mathbf{X}^\mathrm{T}\mathbf{X}$ is not invertible. Its determinant is zero or, equivalently, the matrix is singular.

Multicollinearity refers to cases where at least two columns of $\mathbf{X}$, and hence $\mathbf{X}^T\mathbf{X}$, are highly correlated so that $\mathbf{X}^T\mathbf{X}$ is *nearly* singular. For instance, if we regress house price on the land area and number of rooms, then land area and the number of rooms are likely to be highly correlated, and the estimated parameters will have large standard errors. Generally, we suspect multicollinearity if correlations are higher than 0.8.

An alternative method of detecting the presence of multicollinearity is to compute the *variance inflation factor*

$$\text{VIF}(b_j) = 1/(1 - R_j^2)$$

where R_j^2 is the R^2 obtained by regressing X_j on all other independent variables in the model. If R_j^2 is close to 1, VIF will be large. A VIF value greater than 10 indicates that multicollinearity may be present.

Once we have detected the presence of multicollinearity, there are several remedies, such as:

- leaving things alone because there is no point in trying to estimate more precisely than reality allows;
- redesigning the model to remove highly correlated variables;
- dropping one of the correlated variables; and
- using *ridge regression* (Hoerl and Kennard, 1970).

The first approach assumes that multicollinearity is inherent in the sample. Redesigning the model may improve matters. For instance, if the two independent variables are similarly affected by inflation, then they tend to be highly correlated. A simple redesign of the model that may remove multicollinearity is to deflate both variables by an appropriate inflation index. If there are many independent variables, dropping one of the variables is a good option. If two independent variables are highly correlated, they provide roughly the same amount of information. Hence, we can drop one of them.

Ridge regression adds a small number g to the diagonals of $\mathbf{X}^T\mathbf{X}$ before inverting it to solve the normal equations, that is,

$$\mathbf{b}_R = (\mathbf{X}^T\mathbf{X} + g\mathbf{I})^{-1}\mathbf{X}^T\mathbf{y}$$

where $\mathbf{b}_R$ is the ridge estimator and $\mathbf{I}$ is the identity matrix. We select the value of g so that each element of $\mathbf{b}_R$ appears relatively stable.

The value of g should be as small as possible, that is, it should be just sufficient to stabilize the values of the estimated regression coefficients. This is because the ridge estimator is biased, that is,

$$E[\mathbf{b_R}] = E[(\mathbf{X^T X} + g\mathbf{I})^{-1}\mathbf{X^T y}]$$
$$= E[(\mathbf{X^T X} + g\mathbf{I})^{-1}\mathbf{X^T}(\mathbf{X\beta} + \boldsymbol{\varepsilon})]$$
$$= (\mathbf{X^T X} + g\mathbf{I})^{-1}\mathbf{X^T X\beta} \neq \boldsymbol{\beta}$$

unless $g = 0$, which returns us to the OLS estimator $\mathbf{b}$. Other than its bias, it is also more difficult to derive the sampling distribution of $\mathbf{b_R}$. For these two reasons, the ridge estimator is not as popular as OLS in applied work. We will not discuss it further.

Qualitative Dependent Variable

Sometimes, the *dependent* variable (Y) is binary variable, such as whether a person has lung cancer or whether he is a homeowner or renter. In the cancer case, we can model it as

$$Y = \alpha + \beta X + \lambda G + \theta A + \varepsilon$$

where $Y = 1$ if a person has lung cancer and 0 otherwise. The variable X is the number of cigarettes smoked per day, G is gender, and A is the age of the person. For brevity, we will drop the gender and age variables and consider the simpler model

$$Y = \alpha + \beta X + \varepsilon. \tag{15.12}$$

This is just to save on notation and avoid carrying too many independent variables. The principles are the same if there are more independent variables.

In Fig. 15.6, we plot the data for a sample of eight persons. Five of them have lung cancer (that is, $Y = 1$). For example, John, who smokes four cigarettes a day, has lung cancer. Observe that the dotted regression line fits the data poorly, resulting in low R^2.

A better option is to fit a logistic curve

$$Y = \pi = 1/[1 + \exp(-(\alpha + \beta X))].$$

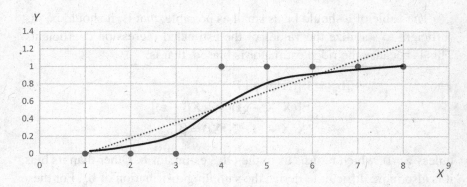

Fig. 15.6   Data on number of cigarettes smoked per day (X) and incidence of cancer (Y).

Here exp(.) is the exponential function. In Fig. 15.6, we show this curve as a solid line. For any X value, we interpret the corresponding Y value on the curve as the probability (π) of having lung cancer.

The logistic curve is a nonlinear function. To transform it into a linear model, we first use simple algebra and rewrite it as

$$\pi/(1 - \pi) = \exp(\alpha + \beta X).$$

The ratio on the left hand side is the odds. Taking logs on both sides gives

$$\log(\pi/(1 - \pi)) = P = \alpha + \beta X.$$

If we add an error term, we can regress P against X using OLS if we have data on P. This means that we need to find data on π, because P depends on π.

To estimate π, we can use grouped data (Table 15.5). Note that we cannot use the first two data points in the table because

$$P = \log(\pi/(1 - \pi)) = \log(0/(1 - 0))$$

is undefined.

Observe that a large sample is required to obtain reliable estimates of π for different categories of smokers. If the sample size is small, we may use the maximum likelihood method discussed earlier. It uses individual rather than grouped data.

Table 15.5 Data for logistic regression.

X	No. of persons	No. with lung cancer	π	P
1	100	0	0	Undefined
2	100	0	0	Undefined
3	100	1	0.01	−1.9956
4	100	2	0.02	−1.6902
5	100	2	0.02	−1.6902
6	100	4	0.04	−1.3802
7	100	6	0.06	−1.1950
8	100	6	0.06	−1.1950

Recall that the likelihood function is

$$L = Prob(Y_1,\ldots, Y_n) = f(Y_1)f(Y_2)\ldots f(Y_n).$$

where

$$f(Y_i) = \pi_i^{Y_i} (1 - \pi_i)^{1-Y_i}.$$

This implies

$$f(Y_i = 1) = \pi_i(1 - \pi_i)^0 = \pi_i$$
$$f(Y_i = 0) = \pi_i^0(1 - \pi_i)^1 = 1 - \pi_i$$

as desired. That is, π_i represents the probability that the ith person who smokes X_i cigarettes a day has lung cancer. Substituting $f(Y_i)$ into L and taking logs gives the log likelihood function

$$Q = \Sigma\, Y_i \log \pi_i + \Sigma\, (1 - Y_i)\log(1 - \pi_i)$$
$$= \Sigma\, Y_i \log[\pi_i/(1 - \pi_i)] + \Sigma \log(1 - \pi_i).$$

Substituting for

$$1 - \pi_i = 1/[1 + \exp(\alpha + \beta X_i)],$$

we have

$$Q = \Sigma\, Y_i(\alpha + \beta X_i) + \Sigma \log[1/[(1 + \exp(\alpha + \beta X_i)].$$

Differentiating Q with respect to the parameters and setting the results to zero gives

$$\partial Q/\partial \alpha = \Sigma\, Y_i - \Sigma\, Z_i/(1 + Z_i) = 0;\text{ and}$$
$$\partial Q/\partial \beta = \Sigma\, X_i Y_i - \Sigma\, X_i Z_i/(1 + Z_i) = 0.$$

Here $Z_i = \exp(\alpha + \beta X_i)$. The solutions to these nonlinear equations are the maximum likelihood estimates.

Since Y_i takes the value of 1 or 0, the coefficient of determination R^2 is lower in logistic regression. Some researchers have proposed alternative measures such as

$$R^2 = 1 - [L(0)/L(\mathbf{b}^*)]^{2/n}.$$

Here $L(0)$ is the value of the likelihood function for the null model without independent variables and $L(\mathbf{b}^*)$ is the value of the likelihood function using the estimated coefficients (see Cox and Snell, 1989).

Reverse Causality

In the linear regression model, we assume that $Y = f(X)$, that is, X affects Y, and not the other way round. We say that X is exogenous or determined outside the system. However, Y may also affect X. If it happens, there is reverse causality and OLS will yield inconsistent estimates.

To see this, suppose we use OLS to estimate the model

$$Y = a + bX + cG.$$

Let us assume that we are interested in the coefficient of G. If Y also affects X, then there is a second equation

$$X = d + eY.$$

If we substitute X into the first equation, we have

$$Y = a + b(d + eY) + cG.$$

After some rearranging, we have,

$$Y = f + gG$$

where

$$f = a/(1 - be); \text{ and}$$

$$g = c/(1 - be).$$

Hence, we have two estimates for the coefficient of G, namely c and g. Our estimates are *inconsistent* in the presence of reverse causality.

As an example of reverse causality, if Y is a person's state of mental health, X is a dummy variable on whether a person is currently working, and G is gender, then X and Y may affect each other. Unemployment can affect a person's mental state and this, in turn, may affect his decision on whether to seek employment.

If reverse causality is present, then we have to estimate the first two equations above as a system. There are many textbooks on how to solve such systems, such as Johnston and DiNardo (1997) and Kline (2015).

Time Series: Autocorrelation

So far, our discussion has primarily been on cross-sectional data. In this section, we consider linear regressions using time series such as

$$Y_t = \alpha + \beta X_t + \varepsilon_t.$$

The subscript has been changed from i to t to reflect that we are dealing with time series. In a time series, the observations are ordered by time, such as monthly observations of rainfall and interest rates. As before, we use the simple regression model to avoid carrying too many independent variables. The principles discussed here can be extended to the multiple regression case. Where it creates special problems, these issues will be highlighted.

There are two recurring problems with time series, namely:

- autocorrelation, where the errors ε_t and ε_{t-h} are correlated for some lag h, which violates the OLS assumption of independence among the error terms; and
- non-stationarity, where a time series is trending, which can result in *spurious regressions*.

A (weakly) stationary series does not trend, that is, its mean and variance remain unchanged over time, that is,

$$E(X_t) = \mu \text{ and } Var(X_t) = \sigma^2 \text{ for all } t.$$

Further,

$$Cov(X_t, X_{t+h}) = \Upsilon_h$$

is also constant for all t. The covariance depends only on the lag (h) and not on the observation period. For example, if we have a long series of monthly interest rates, the covariance at lag 2 (say) is the same irrespective of whether we compute it using data from the 1990s or 2000s.

Most economic time series such as interest rates, housing starts, and imports tend to trend upwards or downwards. They are not stationary.

We will consider the autocorrelation problem first. Usually, we deal with lag one so that ε_t is autocorrelated with ε_{t-1}.

In Fig. 15.7, observe that if the residual is positive in one period, it tends to stay positive for the next period. For example, the residuals for $X = 1, 2,$ and 3 are positive, and the residuals for $X = 4, 5,$ and 6 are negative. Hence, these residuals are not randomly distributed. They are autocorrelated.

Autocorrelation is common in economic time series because the impact of an external shock, such as a spike in oil prices, tends to persist

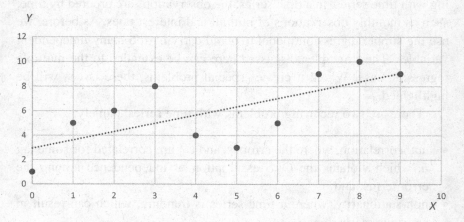

Fig. 15.7 Time series with autocorrelated errors.

for more than one period. Firms and consumers will take time to adjust to the new conditions. At first, higher oil prices will affect industries that use large quantities of oil, such as the oil and transport sectors. Subsequently, the impact spreads to other industries through higher prices for the outputs of these sectors. These higher prices may depress demand and trigger a recession. The process then works in reverse as unemployment rises, investment falls, and so on.

To see the implications of autocorrelation, consider the model

$$Y_t = \alpha + \beta X_t + u_t; \text{ and} \tag{15.13}$$

$$u_t = \rho u_{t-1} + \varepsilon_t. \tag{15.14}$$

Observe that the error term u_t is not random. It correlates with u_{t-1} with ρ as the coefficient of correlation. If ρ is close to 1, the error terms u_t and u_{t-1} are highly correlated. If it is close to zero, they are uncorrelated. The error term ε_t is random and follows the usual OLS assumptions, that is, it has a normal distribution with zero mean and constant variance σ^2. We further assume that u_t and ε_t are independent.

Taking the expectation and variance, we have

$$E(u_t) = \rho E(u_{t-1}) + E(\varepsilon_t) = 0; \text{ and}$$
$$\text{Var}(u_t) = \rho^2 \text{Var}(u_{t-1}) + \text{Var}(\varepsilon_t) = \rho^2 \text{Var}(u_{t-1}) + \sigma^2 > \sigma^2.$$

We can see that the OLS estimator is still unbiased but the coefficient estimates are imprecise, that is, they have larger variances. Consequently, the usual diagnostic t and F tests will not be valid. Hence, we need to test for autocorrelation before regressing time series, that is, we need to test whether $\rho = 0$.

The traditional procedure is to use the Durbin-Watson test but this test may be inconclusive. The modern approach is to use the Lagrange Multiplier (LM) test. The unrestricted model is

$$Y_t = \alpha + \beta_1 X_{1t} + \beta_2 X_{2t} + \rho_1 u_{t-1} + \rho_2 u_{t-1} + \varepsilon_t.$$

The restricted model is

$$Y_t = \alpha + \beta_1 X_{1t} + \beta_2 X_{2t} + \varepsilon_t.$$

Then under $H_0: \rho_1 = \rho_2 = 0$,

$$(\text{RSS}_R - \text{RSS}_U)/d \div \text{RSS}_U/(n - k) \sim F(d, n - k).$$

Here $d = 2$ because there are two restrictions.

If autocorrelation exists, a popular way of estimating the model is to multiply Equation (15.13) by ρ and lagging it by one period to obtain

$$\rho Y_{t-1} = \alpha\rho + \rho\beta X_{t-1} + \rho u_{t-1}.$$

We then subtract this equation from Equation (15.13) and use Equation (15.14) so that

$$Y_t - \rho Y_{t-1} = \alpha(1 - \rho) + \beta(X_t - \rho X_{t-1}) + \varepsilon_t \qquad (15.15)$$

or,

$$y_t = \alpha(1 - \rho) + \beta x_t + \varepsilon_t \qquad (15.16)$$

where

$$y_t = Y_t - \rho Y_{t-1}; \text{ and}$$
$$x_t = X_t - \rho X_{t-1}.$$

We require an estimate of ρ to compute y_t and x_t. Durbin's (1960) method estimates ρ directly by shifting ρY_{t-1} in Equation (15.15) to the right hand side and regressing

$$Y_t = \alpha(1 - \rho) + \rho Y_{t-1} + \beta X_t - \beta\rho X_{t-1} + \varepsilon_t.$$

The coefficient of Y_{t-1} provides an estimate of ρ for use in Equation (15.16) to find y_t and x_t before we can apply OLS.

Time Series: Non-Stationarity and Co-Integration

As discussed earlier, the second problem with time series is that they may not be stationary, that is, both Y and X may be trending upwards or downwards. This is common in economic time series, where it is possible to regress two unrelated series and still obtain a high R^2. We say that the regression is *spurious*. It is a problem because we can never be sure whether the relation between X and Y in our model is real or spurious. Our guide is a theory that postulates how X and Y are related, but it is ultimately only a theory.

A spurious regression is only part of the problem with non-stationary time series. The other problem is that the error variance may be infinite, and this violates the OLS assumption that $Var(\varepsilon_t) = \sigma^2$. To see this, consider the model

$$Y_t = \rho Y_{t-1} + \varepsilon_t.$$

If $\rho = 0$, $Y_t = \varepsilon_t$ and the series is stationary because the random error series is stationary. If $\rho = 1$, we have

$$Y_t = Y_{t-1} + \varepsilon_t.$$

Then, through repeat substitution,

$$Y_2 = Y_1 + \varepsilon_2;$$
$$Y_3 = Y_2 + \varepsilon_3 = Y_1 + \varepsilon_2 + \varepsilon_3; \text{ and}$$
$$Y_4 = Y_3 + \varepsilon_4 = Y_1 + \varepsilon_2 + \varepsilon_3 + \varepsilon_4.$$

Thus,

$$Var(Y_4) = Var(Y_1) + Var(\varepsilon_2) + Var(\varepsilon_3) + Var(\varepsilon_4) = t\sigma^2$$

where $t = 4$. In general, if we repeat the substitution, t will be large and the variance becomes infinite.

In summary, if $\rho < 1$, the series is stationary. If $\rho = 1$, the series is non-stationary. If $\rho > 1$, the series is *explosive* because each Y_t is greater than Y_{t-1}. Hence, the boundary case is when $\rho = 1$, called the *unit root*. A test for stationarity is then a test for unit root.

Consider the model

$$Y_t = \alpha + \rho Y_{t-1} + \varepsilon_t.$$

Based on the discussion above, we want to test

$$H_0: \rho = 1 \text{ (series is non-stationary) against}$$
$$H_1: \rho < 1 \text{ (series is stationary).}$$

Under H_0, both Y_t and Y_{t-1} are non-stationary and, because it violates the OLS assumptions, we cannot use OLS to estimate ρ directly. We need to use the differenced form by subtracting Y_{t-1} from both sides so that

$$Y_t - Y_{t-1} = \alpha + \rho Y_{t-1} - Y_{t-1} + \varepsilon_t.$$

Thus,

$$\Delta Y_t = \alpha + (\rho - 1)Y_{t-1} + \varepsilon_t = \alpha + \delta Y_{t-1} + \varepsilon_t.$$

We call Δ the difference operator, that is,

$$\Delta Y_t = Y_t - Y_{t-1}.$$

Thus, a test of H_0: $\rho = 1$ (series is non-stationary) is the same as testing H_0: $\delta = 0$. This is the Dickey-Fuller (DF) test for unit root. In the Augmented Dickey-Fuller (ADF) test for unit root, we add a more generous lag structure:

$$\Delta Y_t = \alpha + \delta Y_{t-1} + \Sigma \beta_i \Delta Y_{t-i} + \varepsilon_t.$$

The additional lags (ΔY_{t-i}) are included so that the error term becomes random. The asymptotic critical values depend on whether a constant term (α), trend (λt), or both are used (see Table 15.6).

If Y and X are non-stationary series, the usual approach is to regress in differenced form, that is, we regress

$$\Delta Y_t = \alpha + \beta \Delta X_t + \varepsilon_t.$$

This is because a differenced series is often stationary and OLS may be used. For instance, let

$$Y_t = \{1, 3, 4, 5, 7\},$$

which is trending up (i.e. non-stationary). Then

$$\Delta Y_t = \{3 - 1, 4 - 3, 5 - 4, 7 - 5\} = \{2, 1, 1, 2\}$$

has no obvious trend, that is, it is stationary.

However, if we regress in differenced form, we are estimating the short-term dynamic relation between Y and X. More precisely, we are

Table 15.6   Critical values for unit root test.

	Significance level		
	1%	5%	10%
No constant, no trend	−2.58	−1.95	−1.62
Constant, no trend	−3.43	−2.86	−2.57
Constant and trend	−3.96	−3.41	−3.12

Source: Fuller (1976).

estimating how Y changes in response to changes in X. In doing so, we lose sight of the long-term relation between Y and X, which is given by

$$Y_t = \alpha + \beta X_t + \varepsilon_t.$$

Although X and Y are non-stationary, it is possible that the error term is stationary. If this is the case, we say that the two series are *cointegrated*. We may then regress the above long-term relation without worrying about spurious relations, because both series cannot drift "far apart" from each other. Hence, the first step in cointegration analysis is to test whether the residual variable e_t is stationary using the ADF test.

In the second step, we regress

$$\Delta Y_t = \alpha + \beta \Delta X_t + \lambda e_{t-1} + \varepsilon_t.$$

We call the term λe_{t-1} an *error correction mechanism* (ECM). While X and Y have a long-term relation, there may be short-term deviations and the ECM adjusts these deviations towards the long-term equilibrium relation (Engle and Granger, 1987). If there more than two variables, there may be more than one cointegrating relation and more complex procedures are required (Johansen, 1995).

References

Belsley, D., Kuh, E. and Welsch, R. (1980) *Regression diagnostics*. New York: Wiley.

Box, G. and Cox, D. (1964) An analysis of transformations. *Journal of the Royal Statistical Society*, B, **26**(2), 211–243.

Cox, D., and Snell, E. (1989) *Analysis of binary data*. London: Chapman and Hall.

Engle, R. and Granger, C. (1987) Cointegration and error correction representation, estimation, and testing. *Econometrica*, **50**(2), 251–276.

Fuller, W. (1976) *Introduction to statistical time series*. New York: Wiley.

Gruber, M. (1998) *Improving efficiency by shrinkage*. New York: Marcel Dekker.

Hoerl, A. and Kennard, R. (1970) Ridge regression: Biased estimation for non-orthogonal problems. *Technometrics*, **12**, 55–67.

Johansen, S. (1995) *Likelihood-based inference in cointegrated vector autoregressive models*. London: Oxford University Press.

Johnston, J. and DiNardo, J. (1997) *Econometric methods*. New York: McGraw-Hill.

Kline, R. (2015) *Principles and practice of structural equation modeling*. New York: Guildford Press.

Long, S. (1997) *Regression models for categorical and limited dependent variables*. London: Sage.

Maddala, G. (1986) *Econometrics*. New York: McGraw-Hill.

Monahan, J. (2008) *A primer on linear models*. London: CRC Press.

Rao, C. (1965) *Linear statistical inference and its applications*. New York: Wiley.

Searle, S. (1971) *Linear models*. New York: Wiley.

Seber, G. and Lee, A. (2003) *Linear regression analysis*. New York: Wiley.

CHAPTER 16

Concluding Your Study

Format

After the completion of data analysis, the next step of the research process is to develop the conclusion of your study. It consists of the following sections:

- Summary;
- Contributions and implications;
- Limitations;
- Recommendations; and
- Suggestions for future research.

    We will discuss each section below.

Summary

The Summary section, which comprises about two pages, recapitulates the rationale for the research, the research question, the scope of research, and the research objectives. Thereafter, provide a statement of the research framework or hypothesis as appropriate, the research design (including sampling), and the methods of data collection.

    The next paragraph of the Summary section presents the main findings from the data analysis. Do not go into details such as t tests, and so on.

    The Summary section is not the place to introduce anything new, such as a new variable that you have not considered.

Contributions and Implications

This is the "so what?" section. Here, you highlight the theoretical and practical *contributions* of the study. These may include:

* identifying and solving a new problem;
* introducing a fresh perspective to an old problem;
* modifications of an existing theory;
* applying an existing theory to a different environment; and
* discovery of new facts.

Avoid overstating your case. The argument has to be persuasive, but you will lose credibility if you make unreasonable claims.

After discussing your contributions, the next step is to consider the *implications* of your research. The implications may be theoretical or practical, such as the need to modify a public policy.

Limitations

The first limitation of your study may be *philosophical*. For example, you may be interested in causal laws but it may not apply to human behavior.

The second limitation may be theoretical. There may be some theoretical or conceptual problems. For example, cultural theories of economic development cannot explain wwhy certain communities are more innovative.

The third limitation is *methodological*, such as in research design, sampling, and methods of data collection. For example, you may have used a comparative design that does not prove causality or a case study that cannot generalize to other settings. To save cost, you may have used a biased or small sample. Finally, you may have missing or incomplete data.

The final limitation concerns the *analysis of data*. For example, as discussed in Chapter 15, there are many issues in regression analysis, and failure to address these issues adequately may be a limitation of the study. You should anticipate possible objections, and indicate how you have tried to overcome these limitations. Bear in mind that the reader may find your analysis unpersuasive or they may have an alternative theory.

Recommendations

Where appropriate, you may recommend certain actions based on your findings. For example, you may recommend a change in public policy. In this case, you should be clear on:

- the benefits and costs;
- feasibility;
- implementation issues; and
- evaluation.

Suggestions for Further Research

The final section of the concluding chapter may contain suggestions on how to extend the work in future.

A good source of such suggestions is the limitations of your study. Here, you can suggest one or two ways to overcome these limitations.

CHAPTER 17

The Research Report

Format

The format for the research report depends on whether it is a dissertation, thesis, business research report, conference paper, or journal article. For example, a dissertation, thesis, or business research report should include the following:

- Title fly page;
- Title page;
- Letter of transmittal (for business research report);
- Letter of authorization (for business research report);
- Table of contents;
- List of figures and tables (optional), abbreviations (if necessary), and table of court cases (where applicable);
- Acknowledgements;
- Summary (or abstract);
- Body;
- Conclusion;
- Appendix; and
- References.

The *title fly* contains the title of the research report. The title should be concise. Terms such as "A case study," "The relationship between" and "its effects on" are redundant. For example, "Income and housing demand" is a better title than "A regression analysis of the effect of income on housing demand."

The *title page* includes the title of the report, the organization for which the report was prepared, the author(s), and date (see Fig. 17.1).

TITLE

Name
Student number

A dissertation submitted in partial
fulfillment of the requirements for
the Degree of XXX

Department
University
Year

Fig. 17.1   Title page for a dissertation.

NAME OF ORGANISATION
Recipient's name
Position
Organization
Date

Dear XXX,
            Subject
Here is my report on ... which was prepared according to
your letter of authorization dated...

We found that...

The report recommends that...

Thank you for the opportunity to...

Sincerely,
XXX
Name and position

Fig. 17.2   Letter of transmittal.

In a business research report, the *letter of transmittal* releases the
report to the recipient (Fig. 17.2). The *letter of authorization* includes the

person(s) responsible for the project, background, scope, research methodology, fees, deadlines, and other information.

The *acknowledgements* page is not included in a dissertation submitted for examination so that examiners do not know the supervisor(s). The page appears only in the final report.

The *summary* or *abstract* summarizes the research problem, objectives, hypothesis (or framework, as appropriate), methodology, main findings, and implications. This should take about two pages.

As shown in Fig. 17.3, Chapter 1 outlines the research problem, objectives, scope, and organization of study. State the research problem clearly, and explain why it is interesting and important. For instance, the problem may arise from conflicting views about an issue. Then state why its resolution is important.

Chapter 2 provides the literature review and develops the hypothesis or framework for the study. Focus on the key concepts, rather than on the authors. Further, do not review every detail about the problem. As shown in Fig. 17.3, the chapter ends with the hypothesis or research framework.

Chapter 3 provides the methodology:

3 Research Methodology
    3.1 Research design
    3.2 Sampling
        3.2.1 Population
        3.2.2 Sampling frame
        3.2.3 Sampling method
        3.2.4 Sample size
    3.3 Methods of data collection
        3.3.1 Questionnaire
        3.3.2 Pretest
        3.3.3 Interview
        3.3.4 Response rate
    3.4 Data collection and processing
        3.4.1 Selection of interviewers
        3.4.2 Equipment
        3.4.3 Training
        3.4.4 Supervision
        3.4.5 Data processing

CONTENTS

Acknowledgements
Abstract

CHAPTER 1 INTRODUCTION
1.1 Research problem
1.2 Objectives
1.3 Scope
1.4 Organization of study

CHAPTER 2 LITERATURE REVIEW AND HYPOTHESIS
2.1 Definition of housing demand
2.2 Determinants
    2.1.1 Demography
    2.1.2 Income and wealth
    Etc.
2.3 Hypothesis

CHAPTER 3 METHODOLOGY
3.1 Research design
3.2 Sampling
3.3 Methods of data collection
3.4 Data collection and processing

CHAPTER 4 DATA ANALYSIS
4.1 Exploratory data analysis
4.2 Regression analysis

CHAPTER 5 CONCLUSION
5.1 Summary
5.2 Main findings
5.3 Contributions
5.4 Implications
5.5 Limitations
5.6 Suggestions for future work

Appendixes
References

Fig. 17.3 Table of contents.

The exact headings for this chapter depend on the research design adopted. For instance, if you use a case study, then the subsections on sampling are not required. You only need to explain why you select particular cases.

Chapter 4 contains the data analysis. The headings will depend on the type of data you use. If you have qualitative data, the "data analysis" is

Table 17.1   Regression results.

Independent variables	Dependent variables		
	Y	ΔY	$\log(Y)$
Constant	3.63(8.83)***	0.16(2.38)**	0.02(0.82)
T	0.11(3.05)***	—	—
V	1.36(3.54)***	—	—
ΔV	—	1.93(3.19)***	2.31(1.17)
$\log(M)$	—	0.61(6.42)***	0.58(4.91)***
R-square	0.86	0.67	0.54
F-statistic	23.45***	23.24***	7.75*

Figures in brackets refer to t values.
*, **, *** indicate significance at 0.1, 0.05 and 0.01 levels respectively.

largely textual. For instance, if you use content analysis, then the data analyses include the selection, mining, and integration of textual content into a coherent picture. If you use quantitative data, it is advisable to conduct exploratory data analysis to get a sense of the data patterns before applying the more sophisticated statistical methods. Extract, not reproduce, the main items from the printout from statistical software packages (see Table 17.1).

The final chapter of the report contains the *conclusion* and *recommendations*. Use the guidelines given in Chapter 16.

The *appendix* includes charts, detailed calculations, graphs, general tables, data collection forms, maps, and other support material. There is no excuse for sloppy bibliographies, so pay attention to detail.

The Writing Process

Resist the temptation to start writing at the beginning of the research project. Begin with a *brief outline* of the table of contents to provide a structure for the study.

As work progresses, fill the table of contents with *brief notes* on the research problem, scope, and so on. These notes can be in point form. Do not worry if you are not able to specify these items at this stage; they will become clearer as the research progresses. For example, the literature on urban regimes may look like this:

<u>Literature review</u>

Pluralist theory

- Local urban power and decision-making is dispersed (Dahl, 1961)

Elite theory

- Power is concentrated in business community (Hunter, 1953)
- Tends to overstate the power of business

Urban growth machine

- Exerts a high degree of influence through coalition of businesses, real estate, lenders, and possibly labor unions (Molotch, 2007)
- Focuses on local economic development through tax breaks and infrastructure investment to counter urban decline
- Limits to distributional policies because of mobility of capital (Peterson, 1981)

Regime theory

- Cooperation of city's top officials and businesses in *stable* alliance to get things done (Stone, 1989)
- Different types of regimes: Developmental, maintenance, and progressive, with different ideologies, discourses, and strategies

Critique of regime theory

- US-centric and may not exist elsewhere (Keating, 1991)
- Neglect of external influences on cities (Imbroscio, 2010)
- Excessive focus on economic development projects (Pagano and Bowman, 1995)

Write the *first draft* of your research report only after you have completed the data analysis. Thereafter, there will be much rewriting. Do not expect your supervisor to be your proofreader and, even if you write reasonably well, it is helpful to ask someone to proofread your draft for style, logical errors, typographical errors, readability, and so on. Be realistic; proofreading is tedious, and you may wish to ask competent people to proofread only one or two chapters rather than the entire work. The downside is that proofreaders do not get a feel of the entire work.

Writing Style

There are many reference books on writing styles, such as Strunk and White (1979), Day (1998), and Szuchman (2013). What follows is a summary from different sources.

(a) *Voice*

Use the *active voice*. In the past, many journals encouraged the use of the passive voice because it is impersonal. Thus, write

> "I *interviewed* the project manager"

rather than

> "The project manager *was interviewed* by me."

Nowadays, the active voice is increasingly used.

(b) *Tenses*

Use the *past tense* to:

- describe your methodology, for example, "We *interviewed*…"; or
- refer to past studies, for example, "Marx (1886) *argued* that…"

Use the *present tense* to:

- express general truths, for example, "birds *fly*"; or
- discuss your findings, for example, "From the results, I *suggest* that…"

If the time is indefinite, use the *present perfect tense*:

He *has gone* to school.

If there are two actions, use the *past perfect tense*:

If he *had done* what management wanted, they would have promoted him.

The words *did*, *shall*, *will*, and *would* are associated with the present tense:

I *did* not *go* home yesterday.
I *will do* what is required.

(c) *Wordiness*

Cut out unnecessary words by editing it ruthlessly, and never use a long word when a short one will do. For example, the following paragraph is wordy:

Each and every country has its own unique system of registration of contractors. However, the selection criteria that form the basis of the various forms of registration may differ from country to country. The main objective of a system of registration of contractors is to ensure that there are suitable contractors to tender for projects. To tender for projects, contractors should have all the necessary qualifications and be competent in what they are doing...

Better:

The registration of contractors ensures that only qualified contractors tender for projects. Countries differ in their criteria for registration.

Omit needless words. For examples, refer to the list below for some common redundancies (left side) and how they may be shortened (right side):

absolutely essential	essential
add together	add
advance warning	warning
as to whether	whether

at this point in time	now
basic fundamentals	basic
brand new	new
cancel out	cancel
close proximity	proximity
combine into one	combine
combined total	total
complex maze	maze
consensus of opinion	consensus
contributing factor	factor
end result	result
for the purpose of	for
in the event that	if
in the first instance	first
in the majority of the cases	most
investment purpose	investment
lag behind	lag
may or may not	may
mixed together	mixed
new initiative	initiative
new innovation	innovation
occasional irregularity	irregularity
on a weekly basis	weekly
owing to the fact that	since
past experience	experience
personal opinion	opinion
plan ahead	plan
reason why	reason
revert back	revert
serious crisis	crisis
small minority	minority
summarize briefly	summarize
take action	act
the manner in which	how
the reason why is that	because
with the exception of	except

Break *long sentences* into shorter ones, but vary your sentences to avoid monotony. In general, a sentence should not exceed 30 words or about two lines of spacing. A long sentence is hard to read, and may contain too much information or too many ideas for the reader to digest. Rewriting your manuscript is hard work, but remember that wordiness will not attract the intelligent reader.

(d) *Qualifications*

Take a stand when you need to, and qualify your statements if necessary. Do not pepper the report with too many qualifications such as "seem," "apparent," "maybe," "perhaps," "occasional," or "generally." Call a spade a spade; write, "We interviewed 20 shoppers" rather than "20 people" who may not be shoppers.

(e) *Spelling*

Use a consistent style of English, for example, American or British spelling. Non-native speakers tend to mix them up unconsciously, such as film and movie. Here are some examples:

American	British
neighbor	neighbour
center	centre
realize	realise
program	programme
skeptic	sceptic
aluminum	aluminium
gray	grey
kerb	curb
maneuver	manoeuvre
mold	mould
polyethylene	polythene
toward	towards
aging	ageing

Some words have similar meanings:

mutual fund	unit trust
sales tax	value-added tax
amortization	depreciation
receivables	debtors
by-laws	articles of association
real estate	land and buildings

(f) *Metaphors*

A metaphor is a figure of speech. It is a word or phrase that compares two things that are not alike but have something in common.

Metaphors such as "the *heart* of the city" provide useful mental images but they may blind us as well. A city has no "heart" to keep "pumping," whether "healthy" or "ailing." The use of metaphors in writing is inevitable. If properly used, a metaphor serves as an important rhetorical device to persuade the reader. Examples of business metaphors include housing *bubbles*, supply *chain*, demand *curve*, competitive *race*, *lemon* markets, *spiking* prices, *creeping* inflation, and starting on a wrong *footing*.

(g) *Other rules of grammar*

Apostrophes can be confusing for the non-native speaker. In American usage, "James's book" is acceptable.

For quotation marks, Americans use double quotes (e.g. "nice") instead of single quotes ('nice') and place the full stop within quotes, such as:

This dog is "cute."

For a series of terms, there is an optional comma (i.e., the serial/Oxford comma) before "and" such as:

red, blue, and green.

Use "which" to indicate alternatives, for example:

Which unit is your flat?

Otherwise, use "that" as in:

> Beware of dogs *that* bite.

Subject-verb agreements can be tricky. If "one of" is used, use the plural form:

> He is *one of* those *boys* who *are*…

For *each*, *either*, *everyone*, *everybody*, *nobody*, and *someone*, use the singular verb, as in:

> Everybody *loves* Raymond.

For *none*, use a singular verb if it means "no one," as in:

> None of us *is* going.

If we want to mean more than one thing or person, use the plural verb, for example,

> None *are* so capable that *they* can do it by themselves.
> None of us are going.

The word "*any*" may take a singular or plural verb, for example,

> *Any* of the answers is acceptable.

Additional nouns that connect to a singular subject do not change verb agreement, for example:

> James, *as well as* John, *likes* to jog.
> James, *together with* John, *likes* to leave first.

In *collective nouns*, use the singular, as in

> The team *is* strong.

Do not verb nouns. For example, banks lend money, rather than loan money. The variable X has an impact on Y, and not X impacts Y. You conduct experimental trials, but do not trial experiments.

Place adverbs as close as possible to the verb it modifies. There is a difference in meaning between these two sentences:

He *works only* on Sundays.
He *only works* on Sundays.

In the first sentence, he does not work on other days. In the second sentence, he does nothing else on Sundays.

Avoid *sexist or racist language*. Use "supervisor" rather than "foreman." Often, the masculine includes the feminine, that is, use "him" and not "his or her," as in the school regulation "A student should not dye his hair." The plural form is better, for example, "Students should not dye their hair." Avoid "he or she" or the ugly "(s)he."

If abbreviations are used, spell it out the first time, for example, "The Housing and Development Board (HDB)..." If necessary, provide a list of abbreviations.

Writing Form

(a) *Tables, charts and diagrams*

Number tables, charts and diagrams consecutively in each chapter (for example, Figure 4.2 Trends in house prices.). Place table headings at the top or bottom of the table.

Place tables, charts, and diagrams just after (or the next closest possible position to) the paragraph where you refer to them in the text. You may reference them by the expression "As shown in Chart 1.2, ..." or by using brackets, for example, "ABC Corporation relies on internal rather than external financing (see Table 1.2)." Avoid reifications such as "The graph *points to* the fact that..." Graphs do not point, suggest, or show anything.

Use *single line spacing* with a smaller font size (for example, 9 or 10 points) in table cells. As far as possible, do not break tables across pages. If this is not possible, then repeat the Table captions on the next page for truncated tables, for example, "Table 1.2 Output by various industries (continued)".

(b) *Pagination*

Number the preliminary pages (before the first chapter) in lower case Roman numerals (for example, i, ii, iii, and so on) centered at the bottom of the page. The title page is not numbered. Number the body of the report, starting from the first page of the first chapter, in Arabic numbers consecutively.

(c) *Footnotes and endnotes*

Avoid footnotes and endnotes. These side comments, located at the bottom of the page or end of the chapter respectively, tend to disrupt the flow of the narrative.

(d) *Quotations*

Block quotations should be typed single-line spacing and properly referenced by author and page, for example, according to Hije (1980),

> The "organization problem" concerns the nature of labor as a quasi-commodity, one that is *reflexive* and active. Unlike machines, the worker can, to some extent, select the level of effort required. (p. 2) (emphasis added)

Alternatively, place the source after the passage:

> The "organization problem" concerns the nature of labor as a quasi-commodity, one that is *reflexive* and active. Unlike machines, the worker can, to some extent, select the level of effort required. (emphasis added)
>
> (Hije, 1980, p.2)

Do not block shorter quotations, for example:

> As Hije (1980) observed, "those who have trouble with..." (p. 9).

Alternatively, place the page number in front, as in:

As Hije (1980, p. 9) observed, "those who have trouble with..."

Use quotations sparingly. It is better to write the report in your own words. Avoid popular quotations because they look stale. Avoid useless words when citing the literature, for example,

David Hije (1980) in his study noted that...

Here, both "David" and "in his study" are unnecessary.

(e) *Abbreviations*

ibid.	Short form for *ibidem* (in the same place). Often used in conjunction with a footnote where the same source is cited more than once, for example, *Ibid.*, p. 24.
op. cit.	Short form for *operato citato* (in the work cited). Often used in footnotes where we cite the source previously but not immediately preceding, for example, we cite source A, followed by source B, and then A again.
cf.	Compare, as in cf. Hocking (1978).
Chap.	Chapter.
ed.	Editor or edition.
Eds.	Editors.
2nd ed.	Second edition.
et al.	And others, from Latin *et al. ii.* Mostly used in bibliographic references, for example (Smith *et al.*, 1982).
ff.	And in the following pages, as in p. 86ff.
n.	Footnote, as in n.4.
pp.	Pages, as in pp. 34–5.
passim.	In various places in the text.
sic.	So or thus. Inserted in square brackets to show the original source contains an obvious error. For example, "The dog bark [*sic*]..."
v.	Versus. Used in court cases but conventionally translated as "and."
vols.	Volume, as in 4 vols.
Vol.	Volume, as in Volume 4.

(f) *Citations and references*

Various systems of referencing are available. The recommended system for the physical sciences is the number system (for example, [1], [2], and so on). For the social sciences, use the Harvard Referencing System or the American Psychological Association (APA) style.

The author, year and, where appropriate, page numbers may appear in the text in the following manner:

> Garde (1968) found that...However, other studies (Vix, 1974; King, 1978, p. 34) reported...

If two or more publications are cited, the one that is published earlier is cited first (i.e. "Vix" comes before "King" because his paper appeared in 1974, while King's paper appeared in 1978). List the references alphabetically by author's surname at the end of the research report:

> Garde, K. (1968) *Project failures*. London: Wiley.
>
> King, H. (1978) *Project risks*. New York: McGraw-Hill.
>
> Viz, N. (1974) *Project risk management*. New York: Macmillan.

If you cite two or more works of the same author, list them in chronological sequence. For works published in the same year, use Jones (1968a), followed by Jones (1968b).

If there are more than two authors, all the names will appear in the bibliography but "*et al.*" (and others) is used in the text, for example, "(Smith *et al.*, 1983)".

Examples of bibliography

A book by a single author:

> Fox, D. (1969) *Managerial economics*. London: Longman.

A book or technical report by more than one author:

> Strunk, W., and White, E. (1979) *The elements of style* (3rd ed.). New York: Macmillan.

For handbooks, (Vols. 1–3) replaces (3rd ed.); for translated book, (D. Smith, Trans.) replaces (3rd ed.); for technical report, (Report No. 12-1234) replaces (3rd ed.). Indent the second line beginning with "York."

Edited book:

Hall, P. (Ed.) (1966) *Von Thunen's isolated State*. Oxford: Pergamon.

Chapter in an edited book:

Stone, P. (1965) The prices of building sites in Britain. In P. Hall (Ed.), *Land values* (pp. 12–27). London: Heineman.

Journal article:

Pite, D., and Tesa, C. (1981) The crisis of our time. *Journal of Environmental Housing*, **23**(3), 123–141.

Newspaper article, no author:

CIDB perceives strong growth for construction sector. (1993, December 17) *The Straits Times*, p. 47.

Use "pp. 1, 25." for discontinuous article.

Newspaper article, with author:

Tan, T. S. (1993, December 12) URA to auction 12 sites in Jurong. *The Straits Times*, p. 36.

Conference paper:

Unpublished:

Brent, B. (1983, May) *Valuation of hotels*. Paper presented at the meeting of the Society of Valuers, Melbourne, Victoria.

Published:

Brent, B. (1988) Valuation of hotels. In E. Dave (Ed.), *Proceedings of the Third International Symposium on Valuation* (pp. 3–9). Vancouver: Zeti Press.

Unpublished manuscript

Jameson, K. (1993) *Testing concrete strength*. Unpublished manuscript.

Dissertation or thesis

Lim, K. (2000) *Lean building construction*. Unpublished undergraduate dissertation, Department of Building, National University of Singapore.

Replace "undergraduate" with "master" or "doctoral" and "dissertation" with "thesis" where appropriate.

Web site

Global Concrete Network. (2010 Jan). *New understanding of concrete*.

New Jersey: GCN. Retrieved 20 March, 2010 from the World Wide Web: http://gcn.org/concrete.html.

Do not segment the bibliography into journal articles, dissertations, and so on. Place them together.

(g) *Citing legal authorities*

For court cases, the parties to a decision are underlined or indicated in italics but not the connecting "v.," for example, "In *Donoghue* v. *Stevenson* [1932], the House of Lords decided that..." The reference to this case appears in the table of court cases (after the list of figures and tables, not with the bibliography) as:

TABLE OF CASES

Donaldson v. Hemmett (1901) 11 Q.L.J. 35   23
Donoghue v. Stevenson [1932] A.C. 562   28
Etc.

The names of the parties involved are not in italics when they appear in the table. The year of the case is given in square or round brackets according to the status of the law report or journal, followed by the source document, which may be abbreviated (A.C. here, which stands for Appeal Cases, Privy Council and House of Lords, England), but a separate list of abbreviations must appear after the table of court cases. The volume (11) and starting page (35) are also given.

For statutes, brackets may be used in text, for example, "The Chief Surveyor or any Government surveyor authorized by him may undertake field and office checks on the title survey work of a registered surveyor or a licensed corporation or partnership." (Land Surveyors Act, Cap 156, Revised edition, 1991, s. 36(1)). Alternatively, write "The Land Surveyors Act, Cap 156, Revised Edition, 1991, s. 36(1) stipulates that the Chief Surveyor..."

Lawyers often use footnotes in legal citation. For example, "The employee undertakes that he is reasonably skilled,[1] that he will..." appears in the main text and the footnotes are given below.

[1]*Harmer* v. *Cornelius* (1858) 5 CB (NS) 236.
[2]Land Surveyors Act, Cap 156, Revised edition, 1991, s. 36(1).

Sometimes, the footnotes refer only to the section of the Act, for example, in the main text, "The Land Surveyors Act, Cap 156, Revised Edition, 1991[1] stipulates that the Chief Surveyor..."

[1]Section 36(1).

(h) *Units of measurement and numbers*

Use the metric system unless tradition dictates otherwise (for example, per square foot (psf) in quoting property prices). Use the singular form, as in 3cm rather than 3cms, and do not use the period unless cm appears at the end of a sentence. The choice of writing 3cm (no space in the abbreviation) or 3 cm (with space in the abbreviation) depends on the prevailing house style. If the original measurement is in imperial units, the converted metric equivalent appears in brackets, as in 3ft (0.91m). For areas and volumes, use m^2 or m^3 rather than sq m or cu m.

Write out numbers below ten, for example, three houses. Exceptions include ratios (12:1), page numbers, and money ($3). When the number is large, there are several options such as "2.3 million" or "2.3 × 10⁻⁵." Do not begin sentences with numbers; if you must, spell it out, such as "Twenty shoppers jumped the queue." For multiple units such as meters per second, use "m/s" or "ms⁻¹."

Decimals should reflect how precise the item is measured. For example, land areas are expressed to one decimal place (for example, 90.5 m²), and the coefficient of determination R^2 is expressed to two decimal places. Decimals should be consistent, such as 10.2, 23.1 and 0.0 when they appear in tables, and aligned properly.

(i) *Mathematical symbols and equations*

Mathematical symbols are in italics, such as $y = f(x)$. Note f is also in italics but not the brackets. Avoid unnecessarily complex mathematics. If necessary, place long and complicated derivations in an appendix. Letters representing vectors (lower case) and matrices (upper case) are in boldface, for example, **v** and **V**, and not in italics.

Use subscripts to enhance clarity rather than add complexity. Subscripts in expressions such as x_t should be in italics unless they are numerals, for example, x_1. Similarly, superscripts are in italics unless they are numerals. This rule applies to vectors and matrices, for example, $\mathbf{A}_1$ and $\mathbf{A}_i$. The subscripts are not in boldface letters.

All equations are indented or centered and, if required, place equation numbers on the left or right margin, for example,

$$(8.3) \quad R = 2 \log (1 + g) + 3.$$

Refer to it as "Equation (8.3)" or "Eq. (8.3)." Note the full stop after the equation, that is, an equation is part of a sentence. If the equation is in the middle of your sentence, use a comma. For example, the equation is

$$(8.3) \quad R = 2 \log (1 + g) + 3,$$

where R is rent and g is rate of rental growth. Use "log" instead of "ln" to represent natural log because "log n" is clearer than "ln n" or "lnn." Leave a blank space between the operator and operand to enhance clarity. Similarly, note the spacing in "1 + g."

Use Greek letters to represent population parameters and the English alphabet to represent sample estimators. For example, *b* is an estimator of β.

(j) *Layout*

For theses and dissertations, it is common to use 1.5 line spacing for text and single line spacing for table of contents, table entries, and quotations. Use a margin of 25 mm except for the left margin, which is slightly bigger to facilitate binding. Center page numbers near the bottom of the page. Print on both sides of the paper in the interest of sustainability. You should check your university's requirements on the layout.

Illustrations should be neatly drawn and clearly labeled. Where a drawing is complex, use color to indicate different lines. Otherwise, use black printing ink. Enhance the clarity of your work by using diagrams and tables.

References

Day, R. (1998) *How to write and publish a scientific paper*. Westport, Connecticut: Greenwood Press.

Struck, W. and White, E. (1979) *The elements of style*. London: MacMillan.

Szuchman, L. (2013) *Writing with style: APA style made easy*. Belmont, California: Wadsworth.

Appendix

Table A.1 Critical points for chi-square distribution.

	$\alpha(\%)$			
Df	10	5	2.5	1
1	2.71	3.84	5.02	6.63
2	4.61	5.99	7.38	9.21
3	6.25	7.81	9.35	11.34
4	7.78	9.49	11.14	13.28
5	9.24	11.07	12.83	15.09
6	10.64	12.59	14.45	16.81
7	12.02	14.07	16.01	18.48
8	13.36	15.51	17.53	20.09
9	14.68	16.92	19.02	21.67
10	15.99	18.31	20.48	23.21
11	17.28	19.68	21.92	24.72
12	18.55	21.03	23.34	26.22
13	19.81	22.36	24.74	27.69
14	21.06	23.68	26.12	29.14
15	22.31	25.00	27.49	30.58
16	23.54	26.30	28.85	32.00
17	24.77	27.59	30.19	33.41
18	25.99	28.87	31.53	34.81
19	27.20	30.14	32.85	36.19
20	28.41	31.41	34.17	37.57
21	29.62	32.67	35.48	38.93
22	30.81	33.92	36.78	40.29
23	32.01	35.17	38.08	41.64
24	33.20	36.42	39.36	42.98
25	34.38	37.65	40.65	44.31

(Continued)

Table A.1 (*Continued*)

	α(%)			
Df	10	5	2.5	1
26	35.56	38.89	41.92	45.64
27	36.74	40.11	43.19	46.96
28	37.92	41.33	44.46	48.28
29	39.09	42.56	45.72	49.59
30	40.26	43.77	46.98	50.89
40	51.81	55.76	59.34	63.69
50	63.17	67.50	71.42	76.15
60	74.40	79.08	83.30	88.38
70	85.53	90.53	95.02	100.43

Table A.2 Critical points for *t* distribution.

	α(%)			
Df	10	5	2.5	1
1	3.078	6.314	12.706	31.821
2	1.886	2.920	4.303	6.965
3	1.638	2.353	3.182	4.541
4	1.533	2.132	2.776	3.747
5	1.476	2.015	2.571	3.365
6	1.440	1.943	2.447	3.143
7	1.415	1.895	2.365	2.998
8	1.397	1.860	2.306	2.896
9	1.383	1.833	2.262	2.821
10	1.372	1.812	2.228	2.764
11	1.363	1.796	2.201	2.718
12	1.356	1.782	2.179	2.681
13	1.350	1.771	2.160	2.650
14	1.345	1.761	2.145	2.624
15	1.341	1.753	2.131	2.602
16	1.337	1.746	2.120	2.583
17	1.333	1.740	2.110	2.567
18	1.330	1.734	2.101	2.552
19	1.328	1.729	2.093	2.539

(*Continued*)

Table A.2   (*Continued*)

Df	10	5	2.5	1
			$\alpha(\%)$	
20	1.325	1.725	2.086	2.528
21	1.323	1.721	2.080	2.518
22	1.321	1.717	2.074	2.508
23	1.319	1.714	2.069	2.500
24	1.318	1.711	2.064	2.492
25	1.316	1.708	2.060	2.485
26	1.315	1.706	2.056	2.479
27	1.314	1.703	2.052	2.473
28	1.313	1.701	2.048	2.467
29	1.311	1.699	2.045	2.462
30	1.310	1.697	2.042	2.457
∞	1.282	1.645	1.960	2.326

Table A.3   Critical points for F distribution ($\alpha = 5\%$).

n_2	1	2	3	4	5	6	7
				n_1			
1	161.5	199.5	215.7	224.6	230.2	234.0	236.8
2	18.51	19.00	19.16	19.25	19.30	19.33	19.35
3	10.13	9.55	9.28	9.12	9.01	8.94	8.89
4	7.71	6.94	6.59	6.39	6.26	6.16	6.09
5	6.61	5.79	5.41	5.19	5.05	4.95	4.88
6	5.99	5.14	4.76	4.53	4.39	4.28	4.21
7	5.59	4.74	4.35	4.12	3.97	3.87	3.79
8	5.32	4.46	4.07	3.84	3.69	3.58	3.50
9	5.12	4.26	3.86	3.63	3.48	3.37	3.29
10	4.96	4.10	3.71	3.48	3.33	3.22	3.14
11	4.84	3.98	3.59	3.36	3.20	3.09	3.01
12	4.75	3.89	3.49	3.26	3.11	3.00	2.91
13	4.67	3.81	3.41	3.18	3.03	2.92	2.83
14	4.60	3.74	3.34	3.11	2.96	2.85	2.76
15	4.54	3.68	3.29	3.06	2.90	2.79	2.71

(*Continued*)

Table A.3 (*Continued*)

n_2	\multicolumn{7}{c}{n_1}
	1	2	3	4	5	6	7
16	4.49	3.63	3.24	3.01	2.85	2.74	2.66
17	4.45	3.59	3.20	2.96	2.81	2.70	2.61
18	4.41	3.55	3.16	2.93	2.77	2.66	2.58
19	4.38	3.52	3.13	2.90	2.74	2.63	2.54
20	4.35	3.49	3.10	2.87	2.71	2.60	2.51
25	4.24	3.39	2.99	2.76	2.60	2.49	2.40
30	4.17	3.32	2.92	2.69	2.53	2.42	2.33
40	4.04	3.23	2.84	2.61	2.45	2.34	2.25
60	4.00	3.15	2.76	2.53	2.37	2.25	2.17
120	3.92	3.07	2.68	2.45	2.29	2.18	2.09
∞	3.84	3.00	2.60	2.37	2.21	2.10	2.01

n_1: Numerator degrees of freedom.
n_2: Denominator degrees of freedom.

n_2	\multicolumn{7}{c}{n_1}
	8	9	10	20	30	40	∞
1	239	241	242	248	250	251	254
2	19.4	19.4	19.4	19.4	19.5	19.5	19.5
3	8.85	9.81	8.79	8.66	8.62	8.59	8.53
4	6.04	6.00	5.96	5.80	5.75	5.72	5.63
5	4.82	4.77	4.74	4.56	4.50	4.46	4.37
6	4.15	4.10	4.06	3.87	3.81	3.77	3.67
7	3.73	3.68	3.64	3.44	3.38	3.34	3.23
8	3.44	3.39	3.35	3.15	3.08	3.04	2.93
9	3.23	3.18	3.14	2.94	2.86	2.83	2.71
10	3.07	3.02	2.98	2.77	2.70	2.66	2.54
11	2.95	2.90	2.85	2.65	2.57	2.53	2.40
12	2.85	2.80	2.75	2.54	2.47	2.43	2.30
13	2.77	2.71	2.67	2.46	2.38	2.34	2.21
14	2.70	2.65	2.60	2.39	2.31	2.27	2.13
15	2.64	2.59	2.54	2.33	2.25	2.20	2.07

(*Continued*)

Table A.3 (*Continued*)

n_2	n_1 8	9	10	20	30	40	∞
16	2.59	2.54	2.49	2.28	2.19	2.15	2.01
17	2.55	2.49	2.45	2.23	2.15	2.10	1.96
18	2.51	2.46	2.41	2.19	2.11	2.06	1.92
19	2.48	2.42	2.38	2.16	2.07	2.03	1.88
20	2.45	2.39	2.35	2.12	2.04	1.99	1.84
25	2.34	2.28	2.24	2.01	1.92	1.87	1.71
30	2.27	2.21	2.16	1.93	1.84	1.79	1.62
40	2.18	2.12	2.08	1.84	1.74	1.69	1.51
60	2.10	2.04	1.99	1.75	1.65	1.59	1.39
120	2.02	1.96	1.91	1.66	1.55	1.50	1.25
∞	1.94	1.88	1.83	1.57	1.46	1.39	1.00

n_1: Numerator degrees of freedom.
n_2: Denominator degrees of freedom.

Table A.4 Areas under the standard normal distribution.

Z	0.00	0.01	0.02	0.03	0.04	0.05	0.06	0.07	0.08	0.09
0.0	0.000	0.004	0.008	0.012	0.016	0.020	0.024	0.028	0.032	0.036
0.1	0.040	0.044	0.048	0.052	0.056	0.060	0.064	0.068	0.071	0.075
0.2	0.079	0.083	0.087	0.091	0.095	0.099	0.103	0.106	0.110	0.114
0.3	0.118	0.122	0.126	0.129	0.133	0.137	0.141	0.144	0.148	0.152
0.4	0.155	0.159	0.163	0.166	0.170	0.174	0.177	0.181	0.184	0.188
0.5	0.192	0.195	0.199	0.202	0.205	0.209	0.212	0.216	0.219	0.222
0.6	0.226	0.229	0.232	0.236	0.239	0.242	0.245	0.249	0.252	0.255
0.7	0.258	0.261	0.264	0.267	0.270	0.273	0.276	0.279	0.282	0.285
0.8	0.288	0.291	0.294	0.297	0.300	0.302	0.305	0.308	0.311	0.313
0.9	0.316	0.319	0.321	0.324	0.326	0.329	0.332	0.334	0.337	0.339
1.0	0.341	0.344	0.346	0.349	0.351	0.353	0.335	0.358	0.360	0.362
1.1	0.364	0.367	0.369	0.371	0.373	0.375	0.377	0.379	0.381	0.383
1.2	0.385	0.387	0.389	0.391	0.393	0.394	0.396	0.398	0.400	0.402
1.3	0.403	0.405	0.407	0.408	0.410	0.412	0.413	0.415	0.416	0.418
1.4	0.419	0.421	0.422	0.424	0.425	0.427	0.428	0.429	0.431	0.432
1.5	0.433	0.435	0.436	0.437	0.438	0.439	0.441	0.442	0.443	0.444

(*Continued*)

Table A.4   (*Continued*)

Z	0.00	0.01	0.02	0.03	0.04	0.05	0.06	0.07	0.08	0.09
1.6	0.445	0.446	0.447	0.448	0.450	0.451	0.452	0.453	0.454	0.455
1.7	0.455	0.456	0.457	0.458	0.459	0.460	0.461	0.462	0.463	0.463
1.8	0.464	0.465	0.466	0.466	0.467	0.468	0.469	0.469	0.470	0.471
1.9	0.471	0.472	0.473	0.473	0.474	0.474	**0.475**	0.476	0.476	0.477
2.0	0.447	0.478	0.478	0.479	0.479	0.480	0.480	0.481	0.481	0.482
2.1	0.482	0.483	0.483	0.483	0.484	0.484	0.485	0.485	0.485	0.486
2.2	0.486	0.486	0.487	0.487	0.488	0.488	0.488	0.488	0.489	0.489
2.3	0.489	0.490	0.490	0.490	0.490	0.491	0.491	0.491	0.491	0.492
2.4	0.492	0.492	0.492	0.493	0.493	0.493	0.493	0.493	0.493	0.494
2.5	0.494	0.494	0.494	0.494	0.495	0.495	0.495	0.495	0.495	0.495
2.6	0.495	0.496	0.496	0.496	0.496	0.496	0.496	0.496	0.496	0.496
2.7	0.497	0.497	0.497	0.497	0.497	0.497	0.497	0.497	0.497	0.497
2.8	0.497	0.498	0.498	0.498	0.498	0.498	0.498	0.498	0.498	0.498
2.9	0.498	0.498	0.498	0.498	0.498	0.498	0.499	0.499	0.499	0.499
3.0	0.499	0.499	0.499	0.499	0.499	0.499	0.499	0.499	0.499	0.499

The table shows the area under the curve from the center ($Z = 0$) to Z. For example, the area from $Z = 0$ to $Z = 1.96$ is 0.475.

Printed in the United States
By Bookmasters